THE SCIENTIFIC METHOD

T0139378

THE SCIENTIFIC METHOD

Reflections from a Practitioner

Massimiliano Di Ventra

University of California
San Diego, USA

OXFORD
UNIVERSITY PRESS

OXFORD
UNIVERSITY PRESS

Great Clarendon Street, Oxford, OX2 6DP,
United Kingdom

Oxford University Press is a department of the University of Oxford.
It furthers the University's objective of excellence in research, scholarship,
and education by publishing worldwide. Oxford is a registered trade mark of
Oxford University Press in the UK and in certain other countries

Published in the United States of America by Oxford University press
198 Madison Avenue, New York, NY 10016, United States of America

British Library Cataloguing in Publication Data

Data available

Library of Congress Control Number: 2018931028

ISBN 978-0-19-882562-3
DOI 10.1093/oso/9780198825623.001.0001

Printed and bound by
CPI Group (UK) Ltd, Croydon, CR0 4YY

To
Elena, Matteo, and Francesca
in gratitude for their central and unique role in my life

Preface

"OK. Now that I have your attention, let's start!"

I am a theoretical physicist who has been working for over twenty years in academia, publishing regularly in refereed journals, teaching graduate and undergraduate classes, training several graduate students and postgraduate researchers, traveling the world to exchange ideas at conferences, and enjoying the creative aspect of my profession.

Over the years, however, while discussing with educated laypeople interested in Science, students, and even some researchers, I have come to realize that they attribute to Science the ability to answer questions that have always been addressed by other sources of knowledge, in particular, Philosophy or Religion.

With this in mind, I felt the need to discuss these issues with my own students. In particular, I wanted to better convey to them both the *reach* and, most importantly, the *limits* of Science and the knowledge we can acquire from this human enterprise.

Encouraged by their feedback, I offered to teach a class on the scientific method in the Department of Physics of the University of California, San Diego, in the fall of 2012 and spring of 2013.

While researching for my lectures, I realized that the misjudged role of Science is often amplified by the frequent *mis*-representation of scientific discoveries or theories by the media, with us scientists sometimes complicit in this act.

The unfortunate result is that oftentimes—pressed by the obvious desire to increase readership—a scientific hypothesis is elevated to a fact. These "news" spark the imagination of laypeople, sedimenting into our collective consciousness "truths" that have much more in common with science fiction than reality.

Navigating scientific topics may then seem daunting. Not all of us are grounded in advanced Mathematics or Physics or other disciplines necessary to judge the value of a scientific statement.

Even the practitioners of a particular discipline cannot claim expertise in all the other subjects of scientific endeavor, and hence cannot discuss with familiarity all the body of scientific knowledge that is continually generated, seemingly at an ever-increasing rate.

While this working knowledge may not be available to all of us, it is my conviction that it is unnecessary to evaluate a scientific claim. In fact, Science rests on a well-defined *methodology* that provides a guiding set of rules for learning about Nature, and, at the same time, defines the boundaries of scientific inquiry.

Whether we discuss Biology or Chemistry or Physics, the *scientific method* is common to *all* Natural Sciences and is the fabric and the backbone of scientific knowledge. Familiarity with the scientific method is not only a necessary tool in modern times, but it is also well within reach of everyone, whether a practitioner like me, or an interested layman.

This short book at a level accessible to first-year undergraduate students of all Natural Sciences—and quite possibly to high school students—attempts to provide such a tool to a wide audience. It originates from the notes I have written for my class on the same subject.

By Natural Sciences I mean all those disciplines that describe phenomena occurring in Nature, such as Physics, Chemistry, and Biology. (Some of the points discussed in this book may also be relevant to the Social Sciences, such as Economics or Psychology, when empirical approaches are employed.)

Although Mathematics is the language in which we express many of these phenomena, its method and goals are substantially different and will not be treated in this book.

The book addresses the following concepts: objective reality and approximate description of natural phenomena; the role of the observer; the difference among objective facts, hypotheses, and theory; the meaning of "falsifiability"; the "absence of democracy" in the pursuit of scientific truths; and the fundamental and *inalienable* role of experimental evidence in scientific knowledge.

I intend it to be an easy read, and therefore I have made an effort to avoid difficult concepts. The examples I have chosen to clarify the method are taken from well-known and easy-to-follow *facts* so that the reader should easily focus on the method itself rather than the validity of a fact.

My hope is that starting from such examples, by reflecting on the general method, the reader will be able to critically sort through other types of scientific claims, and judge for himself whether they stand on a solid methodology of study or on shaky grounds.

It is, in fact, my belief that a universal understanding of what the methodology of a scientific study is, its reach and limitations, would strengthen our culture, better inform our decisions in related matters, and channel our creativity in building a better future.

Although this is a book on the method of scientific inquiry, I do not intend it to be a treatise on epistemology— "the theory of knowledge." Rather, it is a collection of personal reflections on scientific methodology itself as experienced and used daily by a practitioner.

In this respect, my discussion and understanding of the limits and possibilities of Natural Sciences are very much in line with those of a fellow physicist of the 19th century who wrote extensively about the structure of a physical theory: Pierre Duhem. The famous quote "to save the phenomena," first attributed to Plato, and espoused and extended in Duhem's work, summarizes, in my view, the essence of what a scientific theory should aim for.

Despite this not being a book on epistemology, it will start from, and build upon the philosophical/logical foundations of Science. This starting point is nowadays more important than ever. In fact, by forgetting its foundations, we transform Science into some form of bad "religion" that we may call "Scientism."

Scientism is very detrimental to Science per se and should be rejected unequivocally by all, laypeople and practitioners alike.

The addition of a chapter on this form of "religion," and one on the "What" and "Why" questions, is then my humble attempt to clarify and rediscover the foundations of what it truly means to do Science in an age when the noise of uninterrupted news and fads has a negative impact on our collective knowledge.

My hope is that this book will help rekindle a much-needed interest in the centuries-old structure and foundations of this marvelous human enterprise we call Science.

Massimiliano Di Ventra   La Jolla, 2017

PS: Why the cartoons? You have heard that "a picture is worth a thousand words." I personally think that a humorous picture renders the reading of the present subject lighter. It also conveys the message better and expresses to the reader that we can always discuss serious arguments with a joyful demeanor.

PPS: The book is illustrated by $\mathcal{MD}^2$ (a.k.a. Matteo Di Ventra).

PPPS: "*No one is more hated than he who speaks the Truth.*"—Plato.

$\cdots \infty \cdots$ **PS:** "*My desire is to befriend everybody, but my biggest desire still is to befriend Truth.*"—my own rendering of Aristotle's famous quote on Plato.

Contents

1

Science without Philosophy?

"Didn't I tell you Philosophy is useless?"

A first question to settle before embarking further into the topic of the scientific method is the following.

Can Science stand on its own feet without the "Love of Wisdom," a.k.a. Philosophy?

In an age in which Science has conquered the podium as one of, if not *the* most respected human endeavors, we have been led to believe that we may avoid any philosophical argument to define its method and make it a solid enterprise.

So, we hear statements of the sort "Science is the *only* way we know truth"[1] or "Philosophy is dead."[2]

---

[1] A simple Google search would show that this statement has been written in too many articles to cite. Emphasis is mine.

[2] S. Hawking and L. Mlodinow, *The Grand Design: New Answers to the Ultimate Questions of Life* (Bantam, 2010).

The Scientific Method. Massimiliano Di Ventra, Oxford University Press (2018).
© Massimiliano Di Ventra.
DOI: 10.1093/oso/9780198825623.001.0001

Let us, for instance, analyze the first statement: "Science is the *only* way we know truth." This statement begs the following question: "How can Science prove this statement to be true?"

In no way could we ever perform an experiment that would validate it!

Therefore, the above statement is *not* scientific in itself. It refutes itself, and therefore it is not true.

But then, if *empirical* tools are the only ones at our disposal, how did I arrive at the conclusion that the statement is not true if I cannot prove it scientifically?

I came to that conclusion by means of what we humans innately possess, and which has been formalized from at least the very first philosophers, such as Aristotle: *logical thinking*.

The Logic I am talking about here is the foundation of Philosophy and precedes Logic as a mathematical discipline. It is this type of "philosophical" Logic upon which we build *any* knowledge, whether scientific or otherwise, and we use it every day, often without even realizing it consciously.

For instance, suppose we go to work following a particular road. That road is always dry, but one day we observe it to be wet.

We would then conclude, without actually having a direct proof, that something (e.g., water or some other liquid) must have rendered the road wet. The road cannot become wet if left to its own devices.

Somehow, we know that some other agent, other than the road itself, must have *caused* the road being wet, even though we do not know what that agent is.

We reach that conclusion by implicitly using a form of logical thinking built upon our experience.

If someone asked us how we arrived at such a conclusion, we would simply say that it is *logical* to think that if the road is now wet, when it wasn't the other days I traveled it, something must have made it change its state of dryness. It is just *common sense* to reach that conclusion.

The *reasoning* that goes from the direct observation (the wet road) to claiming that an agent must have caused it to be wet is then simply "logical."

If a friend would tell you that, no, the fact that the road is now wet does not mean anything, namely that its "wetness" is caused by *absolutely nothing at all*—it just is—you certainly would look at him bewildered!

As it is for any reasoning we make in our daily life, we will see that also Science *does* need the logical, foundational aspects of Philosophy to even define its tenets and method. Without such a logical foundation, nothing we do in Science would make any sense.

On the other hand, *not all* logical statements we make are scientific even though they pertain to material things, like the "wet road" example.

In that example, both the cause (water or some other liquid) and the effect (the property of the road being wet) are *natural*. I am not discussing any "supernatural" effects (a ghost showered it?). And, of course, I used my senses (my eyesight or touch) and previous experiences to reach a valid, logical conclusion.

However, despite all this, I never invoked or used the scientific methodology that I will discuss in this book.

I wanted to stress this point, because, in conversations and journal or news articles, we do sometimes encounter examples (and I will provide some in the final chapters) of what *appear* to be "scientific statements" that may be logical per se, but are *not* scientific at all!

These are the statements we need to guard ourselves against, when we go about doing Science or judging the scientific content of some "theories."

Finally, even if we do not discuss material objects, we can still use logical thinking to validate or invalidate a given statement, namely to know its truth value.

For instance, the reasoning that I used to invalidate the statement I discussed at the beginning, although less obvious than the "wet road" example, and not related to our daily experiences, still falls into the category of logical thinking.

In fact, as I will discuss later in more detail, since Science can *only* deal with materially testable, experimentally verifiable statements, the quote "Science is the only way we know truth" is *logically* wrong, because its truth value *cannot* be tested *experimentally*.

And if someone would tell you otherwise, you may also look at them bewildered, because they are behaving, in this case, "illogically."

So, instead of concluding that "Philosophy is dead," we must acknowledge that it is very much alive. As human beings we (should) apply our rational thinking (common sense) in any enterprise we engage in. Science is no exception.

I will only discuss those aspects of Logic that are critical to the scientific method. Before doing this, however, we need to define what the *object of scientific study* is.

Takeaways from this chapter

- Logic is very much essential to do Science and is at the core of its method.

- Statements we make can be logical and yet *not* scientific.

- Science is *not* the only way we know truth. Other intellectual domains, such as Philosophy and Religion, are legitimate sources of knowledge.

- Philosophy is definitely *not* dead.

2

Material World and Objective Reality

"Can't you see it's a cute
little unicorn?"

All scientific disciplines tend to specialize in particular
aspects. Biology studies living organisms, Physics the
temporal and spatial relations between inanimate objects,
and so on.

However, irrespective of the particular subject, all Natural Sciences deal with objects and events that we observe
either *directly* through our senses, such as sight, smell, or
touch, or *indirectly* through their extensions, namely *instruments* that we build to explore objects and events whenever
we cannot probe them directly with our senses.

For instance, certain stars or even planets in our solar
system are difficult, or outright impossible to see with our

The Scientific Method. Massimiliano Di Ventra, Oxford University Press (2018).
© Massimiliano Di Ventra.
DOI: 10.1093/oso/9780198825623.001.0001

naked eyes. Telescopes have been built to precisely expand the reach of our eyes. In general, any instrument we have built to explore our surroundings expands the reach of our human senses.

Irrespective of the subject of a particular scientific discipline we can then safely say that the object of study of the Natural Sciences is the *material world*.

In particular, we study the *relation* between material objects that surround us through our *interaction* with them.

What we observe is a series of *phenomena*, namely what happens to material objects (whether inanimate or living) when they are in interaction with each other and, to some degree, in interaction with the instruments that we use to probe them.

We call all this *material reality* or *material world*.

Here, we can immediately make an aside and say that Science is *only* concerned with the material world. It *cannot* and *should not* venture into anything that does not pertain to the material world.

As we will see in a moment, the *tools* of Science are *empirical*; the *only* information that we can extract from the material reality comes from our interaction with our surroundings.

By "interaction" I mean any direct or indirect action aimed at exploring Nature via our instruments, and by "surroundings" anything that goes from the infinitesimally small (the sub-atomic world) to the extremely large (the Universe we observe).

There may well be a *super*-natural reality. However, the word itself, "supernatural," puts it outside ("supra" from Latin means "above") the natural (material) reality Science can address. Of course, we can, and do deal with it within the confines of Philosophy or Religion, but not Science.

For the sake of Science, we must then limit its reach to the material reality and nothing else.

This is, however, not enough to establish that such a reality is the *same* whether I observe it, or some other person observes it.

Scientists *implicitly* take for granted that this is the case.

For instance, if I and another scientist observe the Moon under the same or similar conditions, we would observe the same object, possibly displaced in space because of its relative motion with respect to Earth, nonetheless still the same object.

In other words, what should we expect of the outcome of several observations of the same phenomenon by different observers but under similar conditions?

Answer: the same result or data, at most corrected for the conditions under which the experiment has been performed. Yesterday, the Moon was in a different position than today, but today's position of the Moon could have been *predicted* yesterday! So, we *agree* that the object of study, the Moon, has not changed, only its relative position with respect to the observer.

It is then *imperative* for Science to declare the following:

Object of study

The *unique material reality* which is *objective*—namely independent of the observer—and which is accessible via our senses or their extensions.

I have encountered people that would deny the existence of such an objective reality. Although this is often the subject of movies or some "beliefs," I cannot truly fathom a scientist claiming the same.

Denying the existence of an objective reality that is present *irrespective* of the observer is equivalent to denying the very existence of Science and its method.

In order to stress how *problematic* such a denial is for Science, imagine if a scientist were to perform an experiment under certain conditions and observe a "reality" different than that observed by another scientist who performs the *same* experiment under the *same* conditions.

Say, one looks into a microscope and observes bacteria, and immediately after another scientist takes his place and observes viruses with the same microscope. These two scientists would come together and exchange the observed data, and find them all different.

Which ones are true? How can they build any understanding of what is being observed?

If that were the case, they would not even be able to agree on what the object of study itself was!

As I will explain later, in a case like the one discussed above, the scientific method would suggest that one of the two, or both, must have made a mistake in their observations, or accidentally replaced the object to observe. Ultimately, all these experiments would have to be repeated until they all agree on the data being collected!

Note that this is also true for quantum phenomena.

In that case, an *individual* measurement of a given physical property may be possibly different for different observers. However, the *collective* outcome of all measurements on the *same* physical property must be the same irrespective of the observer, if the system is probed under the same conditions.

Physicists would then extract from those measurements important parameters, such as the average value of the given quantity under study, or other physical parameters that characterize the phenomenon they are trying to understand. Those parameters are ultimately the ones on which different observers have to agree.

If some substantial discrepancy between the outcomes of measurements performed by, say, two different teams of observers occurs, then something must have gone wrong in one or both experiments.

It would be foolish to think the researchers could conclude that they live in different realities, hence "all is good."

If that were the case, the whole edifice of Science would come falling down, and nothing we do would make any sense!

Note that I am not saying all of them must agree on the *interpretation* of the data, only on the data themselves. We can debate whether viruses are alive, but we should recognize that bacteria are bacteria, not viruses.

I will get back to the role of the experiment more in depth later because it is of fundamental importance. For now, it suffices to agree on the existence of an objective reality.

Now that we have established the existence of such a unique material reality, I am ready to define a point that I have already anticipated.

It is a fundamental tenet of the Natural Sciences. Even more, it is an *operational* tenet, since it defines, *at the outset*, how we should go about doing Science itself. This tenet is its *limit of inquiry*:

Limit of inquiry

The *only* object of study of the Natural Sciences is the material reality. Nothing other than the material reality should be approached by the inquiries of the Natural Sciences, because nothing other than the material world *can be probed experimentally* by our senses or extensions of our senses.

Note that this limit of inquiry is *trespassed* by the previous (false) statement that "Science is the only way we know truth."

The limit of inquiry puts a boundary on what Science *can* and *cannot* access, and declares that the *only* truths that Science has any right to discuss are those that can be probed experimentally, namely *only* those pertaining to the material world.

In other words, there is no point in invoking Natural Sciences to prove or disprove the existence of God, the value of freedom, the meaning of beauty, and anything else that is beyond our experimental reach.

Mind that these are all valid, legitimate, and fundamental questions, without a doubt worth exploring, but with tools that belong to Religion or Philosophy, not experiments, and as such *they do not belong* to the realm of study of the Natural Sciences.

As I have already anticipated, this is also true for *any* statement made by scientists: *if it cannot be tested experimentally it is, at best, a mere hypothesis,* or personal opinion!

As I said, these statements could even be *logical* and yet *not* scientific. To stress this difference, let us go back to the example of the "wet road."

In that example I concluded that *if* I found a road wet, while it was dry before, *then* water or some other liquid must be responsible for its new state of wetness. The road (the material making it) does not have the attributes of being wet if left to its own devices.

So far, I simply reached a valid conclusion using logical thinking. What would it take to make it a scientific argument, namely, one that is testable by experiments?

Never did I ask what *phenomenon* has caused the liquid I observed on the road to be there. *How* did it occur that the liquid happened to be there at that particular point in space (the road) and at that time (the day I observed it)?

In summary, if I ask "How come water was present on that particular road on that particular day?" I am actually asking a very specific and completely different question.

In this case, I am really looking for the natural *phenomenon* that made water fall onto the road and rendered it wet. The question can be addressed in its entirety within the scientific method.

I could say (*hypothesize*) that it rained the day before I traveled that road and, hence, rain is the culprit. I can indeed *test* this hypothesis by looking at the weather record (if it exists) in the area of the road.

If the weather record confirms that it rained that day (at least with some likelihood), *then* I can conclude (with some level of confidence) that my statement ("the rain caused water to fall on that road on that day") is *consistent* with the observation that the road was indeed wet.

Of course, I still used logical thinking in the above example (e.g., in the highlighted "*If . . . then*" statement), but I went a step further by making statements on the possible *phenomenon* that could have produced the observed wetness.

And I have done this all along by working *only* within the limits of inquiry of the Natural Sciences, namely I was only looking for *material* causes, and not some *super-*natural ones.

Takeaways from this chapter

- We access *a unique material reality* via our experimental probes.

- The fact that this reality is unique is a *necessary*, logical premise of doing Science in an *objective* way, namely independently of the observer.

- Science can *only* deal with this objective material reality.

- Science *cannot* deal with subjects or questions that do not lend themselves to experimental tests.

- *All* scientific statements need to be logical, but *not* all logical statements need to be scientific.

- We can reach valid logical conclusions without ever employing the scientific method, namely without ever needing experimental tests of such conclusions.

3

First Principles and Logic

"Excuse me. Is this the
Scuba Diving class?"

Once we have settled on the issue that there is such a thing
as an objective reality—the *only* entity that we can explore
scientifically—we can now ask the question "In which way
do we discover it?"

Unlike animals, we are rational beings and we can let our
imagination roam free. However, when it comes down to
Science we have to recognize that we have quite a few, well-
defined, and *non-negotiable* boundaries that need to be satisfied. I call "boundaries" the constraints of logical thinking
I mentioned before.

We proceed, namely we reason, with a well-defined Logic
that is built in our very nature. This Logic is based on

The Scientific Method. Massimiliano Di Ventra, Oxford University Press (2018).
© Massimiliano Di Ventra.
DOI: 10.1093/oso/9780198825623.001.0001

several important *first principles* or *self-evident truths* (or laws of thought).

The statements I will write down are self-evident because they cannot be proved neither mathematically nor philosophically, nor using Logic itself. In fact, they are its foundations.

However, if I tell them to you, you would certainly recognize them to be true without any doubt or second thought. They just are! In an informal way, they summarize the core of what we call *common sense*.

Anybody who doubts these principles is very limited, if not unable to discuss rationally any argument! Worse than that, try to deny any one of these truths and see if you can even make sense of your own thinking, let alone of a discussion with someone else.

These are the three foundational laws of thought whose formulation is typically attributed first to Aristotle.[1]

The law of identity

An object is the same as itself.

The law of non-contradiction

One cannot say of something that it is and that it is not in the same respect and at the same time.

The law of the excluded middle

For any proposition, either that proposition is true or its negation is true.

---

[1] *Metaphysics* 4.4, W.D. Ross (trans.), GBWW 8, 525–6.

The attentive reader would immediately recognize that I have used them in my previous discussion without mentioning them explicitly. For instance, when I said that any one of those bacteria observed is a bacterium—they are what they are—I made use of the "law of identity."

The "law of non-contradiction" is the statement that *truth cannot be self-contradictory* because something cannot be *both* true *and* false at the same time and in the same way.

For instance, bacteria cannot be bacteria *and* viruses at the same time and in the same way. Or if a scientist says he observes a cat, either that cat is alive or its negation is true ("the cat is dead"). He cannot claim the cat is *both* dead *and* alive at the same time and in the same way.[2]

You should have also immediately recognized that the sentence I wrote before, "Science is the *only* way we know truth," is false since its truth value (whether it is true or false) has no means of being proven only from its own meaning (namely that *only* Science is able to find truths).

In other words, that statement is a contradiction because it excludes the possibility of another one ("there is some *other*, non-scientific way of knowing the truth"), and yet it claims to be true: if there is no other way of knowing the truth other than Science, how do I know this statement is true, since the statement itself is not a scientific one and yet claimed to be true?

---

[2] Quantum Mechanics *does not* contradict this law (or any other law of thought for that matter), as some news articles may seem to imply. The famous Schrödinger's cat gedanken experiment acknowledges that when the cat is *observed* it is either dead or alive, not in a superposition of dead and alive states. I'll come back to Quantum Mechanics and its apparent issues with these laws of thought sometime later.

This is, indeed, an application of the "law of excluded middle," which says either that a statement is true or its negation is true.

To these three first principles we still need to add another one, which, in fact, I have already implicitly used.

Suppose I see a ball rolling on the floor but I do not see what made it move. Even if I do not see what caused such movement, I know *for sure* that the *cause* of such an *effect* (the ball rolling) must have been something or someone pushing it somehow.

In other words, we can stipulate the law of cause and effect:

The law of cause and effect

For an event to occur at a given time, there must be another event that has caused it.

There are actually two ways to look at this principle: one philosophical, one scientific.

In Philosophy we can abstract the two entities of this law (cause and effect) and consider logical conclusions that have nothing to do with Science.

As an example, one could start from the observation of motion of an object (e.g., the ball rolling) and conclude that some cause must have generated such motion (e.g., a child kicking the ball). But, in turn, that particular cause can be viewed as the effect of some other cause, and so on.

Strictly speaking, in Philosophy, a "causal regression" (whether the one I discussed above or any other) does not need to be confined to a specific set of objects or phenomena. In fact, those causes could be *anything* in the Universe, and affect *anything*, ultimately even *any* notion, e.g., space and time, or our own thoughts.

An important consequence of all this, which we may want to mention now, pertains to the philosophical meaning of "creation."

In fact, if we abstract this "causal regression" to *anything*, either it may never end or we can subscribe to the notion that there exists a "First Mover," who is the *ultimate cause* and is "self-sufficient," namely *does not* need to be caused.

We typically call this "First Mover," God. He would have put into *existence* (hence in motion) the *whole* reality, *including* space, time, matter, etc., from the *true nothing*, the philosophical concept of absence of *anything* that exists.

Philosophically, this is creation *ex nihilo*, precisely because it comes from the *absence of anything*.

On the other hand, let's not confuse this "absence of anything" with the "vacuum" that is discussed in Quantum (Field) Theory. The quantum vacuum is a *physical* entity with physical attributes of space, time, energy, etc.

When we say "absence of anything" in Philosophy, we really mean *everything*, including absence of those attributes!

Note, however, that although I have used logical thinking to discuss the presence or absence of a "first cause," the conclusions I have reached are within the realm of Logic (or Philosophy in general), but are definitely *not* scientific.

Why? Because, while I may *logically* reach the conclusion that there exists a *super*natural being (or even a "supernatural reality"), *outside* even space and time, the very fact that it is *super*natural puts it *outside* of the limit of inquiry of the Natural Sciences.

In simple terms: there is no way to *experimentally test*, with any possible physical instrument, such a reality, even if it existed.

Now that we have briefly discussed the reach of the law of cause and effect in a philosophical context, let's move on to its meaning in Science.

In a scientific sense, we use the law of cause and effect (causality) in a much more *restrictive* way.

Rather than considering its "abstract" or philosophical conclusions, we employ it in a more *practical* way.

So, what qualifies as "cause" of a given event, in a scientific sense, and when should we be content that we have found it?

This is a *very* important point to consider since it defines the boundaries of what we can say about a given phenomenon or a set of phenomena.

In fact, in Science, answering the "cause question" of a phenomenon is tantamount to deciding how we approach it, in terms of experiments to perform, and the theoretical description we provide of such a phenomenon.

In other words, the scientific "cause" of an effect is related to our ability to measure (test) that effect; for example, how far our measuring instruments can go in observing the natural world, as well as the level of description we, scientists, are willing (or able) to provide of such a phenomenon.

To make things clearer take as example the rolling ball. I may discover that it was kicked by a child, and I could be satisfied with this plausible cause ("explanation").

However, I could be more *descriptive* and say that the child's foot exerted a force and imparted a change in momentum on the ball, which then moved.

Yet more: the atoms of the child's foot, collectively, through their mutual interaction, received energy from the rest of the leg and moved at a given speed. These atoms interacted with the atoms making up the ball, and through repulsive interactions they transferred momentum to them. Or the quarks of the nuclei of the atoms.... I think you get the point.

This exercise needs to stop at some point. At which point it stops is determined by both the investigator himself and, to a certain degree, the instrument of investigation. The investigator may not be satisfied with a particular description of an effect and may then advance a new hypothesis of what may have caused a particular effect, but the instruments at his disposal to test such hypothesis are not available.

Following the above example, we could say that the ball moving is the result of the atoms in the leg of the child, but our naked eyes have no means of testing such a statement. Therefore, if our eyes are the only observational instrument at our disposal, we can be content with the *description* ("explanation") that the child's leg has imparted momentum to the ball, without worrying about the atoms in his leg.

Of course, this does not preclude a researcher from making *predictions* about a more "elementary cause" of that particular phenomenon, even though the instruments at our disposal do not allow a test of such a prediction as of yet.

Therefore, even if the eyes are the only instruments at our disposal, the researcher could still argue that there is a more "elementary cause" behind the momentum imparted by the child's foot, and posit the existence of atomic entities in the child's leg.

This researcher's *description* of what caused the phenomenon would then be considered more "refined" or "elementary."

However, as I will explain in depth later, to qualify as scientific, this description must lead to *testable predictions* that could come to fruition—read, be *tested experimentally*—when new instruments and experiments have been devised precisely to test such predictions.

Otherwise, this description will simply remain in the realm of hypothesis, or worse, be fruit of the researcher's imagination.

In conclusion, in Science the *level of description* of an event and its "cause" are not abstract concepts *unrelated* to the observer(s). They are determined by the observers' degree to which they approach that particular natural phenomenon and the experimental probes at their disposal.

Note that, due to the limit of inquiry I discussed earlier, it follows that the cause of a given effect must be *natural*. As scientists, we are not allowed to invoke "supernatural" causes—a ghost, a miracle—namely causes that are outside the material reality , which is the subject of study of Natural Sciences.

We may, in addition, clarify that causality is distinct from *correlation*. The latter simply establishes a relation between two events. For example, when I saw the ball rolling, I also heard a loud noise. The two events may be correlated, but not necessarily one follows from the other.

Hopefully, all this discussion should have elucidated even further the difference between *scientific* statements and *logical* ones.

Even though both need to build upon the foundations of Logic, the latter ones do not need to correspond to a scientifically testable material reality.[3]

I will exemplify these issues further in the book's last chapters because very often they create confusion on what Science can and cannot address. For now, let us move on and stress even more the fundamental role of experiments.

Takeaways from this chapter

- We use the foundational laws of Logic (first principles) to think rationally and do Science.

- The application of the law of cause and effect is somewhat different in Philosophy from that in Science.

- The philosophical "vacuum" is different from the physical "vacuum." The former is absence of *anything*, including physical attributes. It *cannot* be measured. The latter has physical properties that are measurable.

- The scientific "cause" of an effect is related to *both* our ability to measure the phenomenon we are interested in *and* the level of description we decide to use at the outset.

---

[3] The majority of—and in a very strict, formal sense, *all*—statements in Mathematics are like this. Mathematical statements only need to follow logically and unequivocally from the foundational axioms of Mathematics, regardless of whether they correspond to a material reality.

- This description, however, needs to lead to *testable predictions* to be considered a valid, scientific description.

- Otherwise, it remains within the realm of hypothesis or simply fruit of our imagination.

4

Natural Phenomena and the Primacy of Experiment

"Those are the referees that asked him to test his theory on more than a stone and a feather."

We have now established that the object of study of the Natural Sciences is the (unique) material reality around us. We, the observers, are able to explore such a reality with our senses or extensions of our senses by probing the inter-actions between the material objects and our instruments, interactions that we call *natural phenomena*.

For instance, when I look at two objects colliding, my eyes collect the light that bounces off those objects for the whole duration of my observation. My brain then analyzes that information to make sense of what I see. Without the interaction of that light with my eyes, I would not be able to see the objects at all. The same goes for any other type of measurement we make of phenomena.

The Scientific Method. Massimiliano Di Ventra, Oxford University Press (2018).
© Massimiliano Di Ventra.
DOI: 10.1093/oso/9780198825623.001.0001

However, what part of the material reality do we actually explore?

First of all, we need to decide what type of phenomenon we are interested in.

Maybe we are interested in the motion of the planets in our solar system. Or we are interested in how a certain organism interacts with its immediate environment or some other organisms.

Irrespective, our goal is *not* to study the *entire* material reality at once. Indeed, that would be an impossible proposition. We are limited beings and our instruments have a limited reach.

Therefore, we *limit* our inquiries to an extremely small subset of such a reality. In fact, as we discussed in the preceding chapter, we are limited by the instruments we use to probe it, since the latter ones determine not just the type of phenomena we can address, but also, for a given phenomenon, the type of information we can extract from our measurements.

We can also reverse this argument: as scientists, the *only* aspects of the material reality we can explore are those within reach of our *experiments*. To be precise, we can *only* explore what our senses and extension of our senses (our measurement apparati) allow us to do.

Of course, by inventing new instruments we expand the reach of possible phenomena we can study. Hence, we can interrogate more of the material world.

However, anything that does *not* result from our measurements of the material world is either a hypothesis or pure speculation.

Despite what we may sometimes hear, unless we are able to *experimentally* investigate the world around us, we, as scientists, can only offer hypothesis or opinions ("educated guesses"?), which cannot contribute to the body of knowledge we call Science.

This shows why experiments constitute both the *starting point* and the *end goal* of the Natural Sciences.

Now that we have realized that we can probe, at any given time, only a *limited* aspect of the material world with our instruments, we can discuss how we go about doing that.

Ideally, we would like to keep the interaction of the measurement apparatus with the material objects we want to study as small as possible so that the apparatus does not perturb considerably what is happening to those objects. Of course, this interaction cannot always be made small, but at least we should be able to know the extent of its perturbation on what we observe.

If we know how much the probes influence the measurement, by repeating the experiment many times, even with different types of probes, we should be able to *reproduce* the same results *within the measurement errors*.

For instance, if I want to measure the distance between me and an object, I can measure such a distance directly using a ruler, or, if that is impractical because the distance is quite large, by sending radio waves (using a radar) that bounce off the object and return to my position after some elapsed time. Through the knowledge of the speed of radio waves in air I can then determine such distance.

Regardless of the measurement process used, to claim that such a distance is, say, a kilometer, I need to repeat such a measurement many times. Each measurement will be slightly different from the others (say, one kilometer plus or minus a few meters) due to factors that are difficult to control. (For example, in the case of the radio waves, possible perturbations from the air density variations in between me and the object.)

However, if I repeat the measurement many times my confidence in the fact that the distance is one kilometer increases, and I can ultimately assign that value *averaged* over all the measurements, plus or minus some of its fraction (the measurement error or uncertainty).

If I were able to measure such a distance with a ruler, then such type of measurement would again need to be repeated many times, producing an average value of the distance and the associated error. By comparing the two types of measurements (one with the ruler, the other with the radar) I can say that the two are *consistent* (or that "they agree") if the average of one and the average of the other are within errors from each other. For instance, the measurement with the radar gave me a distance of (1000 ± 1) m, while the ruler one produced (999 ± 2) m. Then the two measurements are consistent.

If, instead, the ruler measurement gave me (990 ± 2) m then, even if I take into account the error in this measurement, its maximum value would be 992 m, while the radar measurement would be, at a minimum, 999 m which is still 7 m larger than the previous maximum value. The two measurements are *inconsistent* (they "*do not* agree") within their respective errors.

If the two measurements are inconsistent, we may perform a third type of measurement with different instruments (e.g., by walking all the way with equal steps; tough but doable), or we realize that one of the two (or both) had, in fact, some systematic error that we did not detect at the outset. For instance, the measurement with the ruler was skewed by our inability to accurately reset its position at every step of the measurement, thus giving an effective larger error than previously estimated.

Regardless, a given phenomenon is declared to be represented by *objective data* or *facts* when *measurements* are performed, and a *consistent* (average) value can be assigned with a given error to its physical properties (the distance in the previous example). We then define objective data or facts as follows:

Objective data or facts

Those obtained by *consistent* measurements on the same phenomenon, *irrespective* of the observer.

How many times should a measurement be performed? That depends on the type of measurement one performs and the desired level of confidence (error) we are willing to tolerate. It may be impractical to walk one kilometer with well-defined steps many times over, but if we are happy with an error of, say, 10 meters we could probably get away with just three strolls.[1]

---

[1] The "error analysis" I have presented here is very simplistic. It is not intended to capture all the subtleties of the measurement process. A more in-depth discussion of measurement and error analysis can be found in, e.g., the book by J. R. Taylor, *An Introduction to Error Analysis: The Study of Uncertainties in Physical Measurements* (University Science Books, 1996).

The data are then declared *objective* because another observer could *repeat* the experiment, maybe with yet another type of instrument, and confirm the previously recorded results, possibly with a smaller uncertainty.

If this observer finds inconsistent data with respect to what has been done previously, then the process needs to restart: either these new data are somehow wrong or they are affected by some other effect not taken into account or nonexistent in the original measurement.

Alternatively, the new data may have been able to uncover a new aspect of the phenomenon not previously observed because the previous instruments could not reach that level of accuracy in studying that phenomenon.

Incidentally, this is how many new discoveries in Science are made: by building new instruments that can reach farther than the previous ones. A case in point is the telescope built by Galileo that allowed him to see, for instance, features of the Moon and Saturn's rings previously unseen by naked eyes.

As an aside, the above procedure is true also for Quantum Mechanics and, indeed, by keeping in mind the experimental way in which quantum phenomena are probed may dispel much of the "myths" surrounding them.

For instance, when we say that a quantum mechanical entity is "*both* a particle *and* a wave" (as if we could contradict the "law of identity") we are actually misinterpreting the experimental results quite grossly.

To be more specific, take the famous two-slit experiment in which a beam of particles, say electrons, impinges on a screen with two narrow openings. On the other side of the screen there is yet another one (call it screen 2) that collects the particles that pass through the slits and measures their position. Every time a *single* electron hits screen 2, one observes a dot where the collision occurs, indicating that the electron behaves as a single particle at that particular position.

However, once *many* electrons have hit screen 2, by looking at it somewhat from afar, we see an interference pattern typical of waves. It is *not* the *single* electron that shows wave properties but *all of them* together!

In fact, in order to see the wave properties we are forced to measure *many* electrons colliding on screen 2, which is precisely what the scientific method forces us to do anyway: a *single measurement* on a *single* electron would not show the "wave" behavior!

Note that, at this point in time, we do not know if there is another cause responsible for the observation of such an interference pattern of atomic particles (the famous "non-locality" of Quantum Mechanics): we simply observe it and *describe* it quite well with the Quantum Mechanics equations.

In other words, we have not been able yet to "regress" one step back in the "law of cause and effect" and probe some other phenomenon (some other particles or fields that pervade the Universe?) that could be the cause for this effect.

Will we ever find such a cause? The scientific successes of the past give us much hope for the future, but the real answer is: nobody knows!

Irrespective, this chapter should have conveyed an important point worth stressing once more. The information we collect of the natural world (the objective data or facts obtained by consistent measurements and independently of the observer) is *limited* by how "far" our instruments can "see."

By improving our experimental probes or inventing new ones, we can see farther, towards either the tiniest scales of the atomic world or the farthest corners of the Universe. This allows us to discover new phenomena and to make more sense of others that we have already observed (such as, possibly, the interference pattern of atomic particles).

Since our interest and our measurements *always* pertain to just a particular (extremely small) subset of the whole natural world (say, the electrons impinging on a screen, but not their interaction with the surrounding light), that particular subset of material reality is very often difficult to isolate completely from the rest.

As such, the information we extract from our measurements *always* carries a level of uncertainty.

Or to put it differently, there is no such a thing as "absolute certainty" in the knowledge of the material world that we acquire experimentally. This is true for "classical" or "quantum" phenomena. It is simply an intrinsic feature of the measurement process, and applies to all Natural Sciences with no exception.

Takeaways from this chapter

- Our instruments can only access a *limited* subset of the material world at once.

- Our measurements *always* carry errors.

- Data about a given phenomenon are objective when obtained by *consistent* measurements on the same phenomenon, *irrespective* of the observer.

- *Consistent* means that the data from different observers agree within their respective measurement errors.

- *Inconsistent* data among researchers require either a re-evaluation of those data or additional measurements, so that, ultimately, all data are reproducible within measurement errors.

5

Observation and Experimentation

"Controlled experiments have shown
that scientists *always* discuss
their results in a friendly manner."

Let me linger a bit longer on the concept of *repeating* an experiment for the benefit of obtaining objective data.

As I have already mentioned, in order to explore the material reality around us we need to interact with it either with our senses or with instruments. The object of study is then never truly independent of the way in which we observe it. With this, I mean not just the instrument used to probe the phenomenon, but also the environment in which I perform the measurement.

For instance, if I want to test the classical law of dynamics which says that the force exerted on an object is proportional to its acceleration (variation of velocity in time),

The Scientific Method. Massimiliano Di Ventra, Oxford University Press (2018).
© Massimiliano Di Ventra.
DOI: 10.1093/oso/9780198825623.001.0001

I may track the trajectory of a ball falling under the gravitational pull of the Earth. In doing so, however, I cannot eliminate completely the air surrounding it. The air exerts a drag on the ball that is proportional to its velocity, not its acceleration. This effect is even more evident if you let a feather drop: it will swing slowly toward the ground.

In fact, if I got a ball with the same mass as the feather and would let them both fall at the same time and from the same height with respect to the ground, the second law of dynamics would predict that the ball and the feather reach the ground simultaneously. This is in obvious contradiction with the observation that the ball reaches the ground while the feather is still rocking in midair!

If I make the statement (hypothesis) that the air is responsible for the disagreement, I can come up with an experiment in which I place the ball and the feather in a sealed chamber where I can take out as much air as possible with a vacuum pump, and then repeat the above experiment and check the results. If the air vacuum inside the chamber is really good, meaning very little air remains in the chamber, the experiment will show that the ball and the feather indeed reach the ground essentially at the same time (within measurement errors).

As you can easily see, this type of operation constitutes a "controlled" *laboratory experiment* where the environment (the air) with which my objects of study (the ball and the feather with the gravitational pull of the Earth) interact is well under control. My measurement can take into account not just the error of my instrument (say my eyes). Also the effect of the environment can be accounted for.

However, such controlled laboratory experiments cannot always be performed. For instance, in Cosmology, it is quite difficult to test directly an individual hypothesis or eliminate the role of the environment, or sometimes even repeat the experiment on the same object under the same conditions. The reason is simply because the object under study does move with respect to Earth, and it will not return to the exact initial condition of the experiment in any foreseeable future.

Nevertheless, also these "one-time" observations are facts that contribute to constructing an economical, logical synthesis that attempts to make sense of apparently diverse and distinct effects. Such interpretation is what we call a "theory."

I will discuss later the role of a theory. But first, let us discuss in the next chapter an important aspect that makes the whole Science enterprise work. This aspect relates to how we, scientists, use the work of our fellow scientists.

Takeaways from this chapter

- We can *never* completely isolate the particular phenomenon we are interested in from the influence of its material environment.

- Controlled laboratory experiments on a particular phenomenon may not always be easy to perform.

- This means that we cannot always completely know the effect of the environment on the particular phenomenon we are interested in.

6

The Role of Human Faith in Science

"Which part of 'TRUST ME!'
do you not understand?"

We know that there are eight planets in the solar system; that light travels in vacuum at about 300,000 km/sec irrespective of the relative speed of the source and the observer; that the ionosphere (the upper layer of our atmosphere) reflects radio waves, enabling us to communicate over large distances of the globe; that atoms support discrete energy levels, and so on.

How do we know all this? I personally never did any experiment to test any such claims. However, presumably, someone did, some time ago, and others have repeated them. These statements were then declared objective data or facts.

The Scientific Method. Massimiliano Di Ventra, Oxford University Press (2018).
© Massimiliano Di Ventra.
DOI: 10.1093/oso/9780198825623.001.0001

We read them in some textbook or journal article, and, importantly, we *trust* the sources of such claims. The word "trust" is the correct one here: we indeed have *faith* in what scientists have done in the past (and are currently doing) and we do not question the data that they have generated once these have been declared objective, namely reproduced by other researchers.

This type of *human faith* (as opposed to *supernatural faith* in God) is an important prerequisite to make the whole thing work: it would simply be unrealistic to repeat all the experiments ever done since the beginning of humanity on anything that pertains to the Sciences.

We simply accept that if some honest, accurate, and reproducible work has been put into performing such experiments, then they belong to the collective body of knowledge.

(When I was studying mathematical analysis as an undergraduate, our professor made a somewhat similar statement regarding Mathematics: no one can ever *re*-demonstrate *all* theorems. It is enough that for a given theorem you know *at least* one person who has demonstrated it. I think he was joking, but the statement holds some truth.)

Of course, it could happen that by repeating the same experiments, with more precise instruments, new phenomena may be uncovered.[1]

Once they have been declared objective, as dictated by the scientific method, they will also enter into our accepted,

[1] Or what we thought were objective data turn out not to be the case because of faulty experiments.

collective knowledge, and *faith* in their validity is our only way to move on to other discoveries.

This type of faith, however, need not be the same as "faith" in the *interpretation* of the objective data.

In fact, I need to stress here (and I will expand it later as well) the important difference between the objective data (facts) themselves and how we *interpret* those data. The interpretation of data is not necessarily objective, namely independent of the observer.

For a given set of data, there could be different ways in which scientists make sense of them.

Take, for instance, the attraction of a body to Earth. What we call "attraction" is a fact, very much in plain sight of anyone on this planet.

However, we usually go a step further and say that this attraction is due to a "gravitational force."

Despite being part of the lexicon of everyone, starting from elementary school, the name "gravitational force" is simply an *interpretation* of the fact that bodies are attracted to Earth (and attract each other anywhere in the Universe).

Why am I calling it an "interpretation?"

Isn't it a "fact" that the gravitational force is responsible for the attraction of any massive object to any other one, irrespective of whether these objects are confined to Earth, or anywhere in the Universe?

It all depends on what we mean by "fact" here.

If by "fact" we mean what we have just discussed, namely that objects attract each other, then we all agree that this statement is correct. For instance, it has been verified repeatedly, on objects of different material properties, shapes, size, etc.

Then I can say that instead of calling it "attraction," I have simply given it a new name, "gravitational force," and we should understand each other on what we mean. All is good.

However, if by "fact" we mean that such an attraction is *described* by a *well-defined* (mathematical) concept we call "gravitational force," then there is room for debate.

In fact, the concept of "gravitational force," namely a vector (like an arrow with a direction and amplitude indicating where the attraction points to and how strong it is), exists only in classical Newtonian Mechanics, and with appropriate mathematical modifications, also in Einstein's Special Relativity.

However, it is just a *mathematical description* of the attraction we have just discussed. Therefore, this particular description can change, and indeed it has *already* changed to some extent!

For instance, in the theory of General Relativity, the attraction among objects is *not* described by a vector we call "gravitational force."

Rather, it can be *described* and *interpreted* as originating from the deformation of the spacetime continuum due to the presence of masses. Two objects attract each other by traveling the shortest path in this curved spacetime continuum they themselves perturb.

Therefore, this example shows again that there is a *fundamental* and important difference between the *facts* we observe (e.g., the attraction between objects) and the *description* (or *interpretation*) of such facts (e.g., using the concept of a "force" or "bending of spacetime").[2]

Although the collection of objective data may sometimes generate some procedural discussions among scientists (e.g., how many measurements have been performed in a particular experiment, on how many different objects sharing the same properties, etc.), oftentimes it is the *interpretation* of such data that creates many heated discussions and animosities in the scientific community.

Ultimately, however, experiments (should) settle any dispute. This last important point deserves much more discussion than a single sentence. For this, I will devote more space in the coming chapters where I will explain what a scientific theory is.

Before doing that, though, let us point out another limit of scientific inquiry.

Takeaways from this chapter

- We must rely on, and have *faith* in the data collected by other scientists, if these data have been found to be objective.

- This is because there is *no way* for us to *repeat* all the possible experiments that have been performed from the dawn of humanity.

[2] One could argue that even the word "attraction" is descriptive. Indeed, we need some "basic language" to communicate with each other and to point to the phenomena we want to study. This "basic communication language" is, however, not usually the type of "description" that I mean here or in the rest of the book.

- This type of "trust" in our fellow humans is, however, *not* the same as faith in the supernatural, e.g., God.

- The trust we put in the collection of data does *not* translate into "faith" in the *interpretation* of those data.

- The same data can be interpreted differently.

- Hence, *interpretation* of facts is *not* necessarily objective, namely independent of the observer.

7

Approximate and Limited Description of Natural Phenomena

"I think your theory may be a bit over-reaching."

In view of our inability to study *at once* the *entire* material reality, our experiments necessarily address *only* a very small subset of such a reality.

To this constraint, which I may call an "experimental limitation," we need to discuss yet another limitation, which is related to the *description* of such a probed reality. I call this extra constraint a "description limitation." To explain this point let me again start with an example.

If I observe a ball falling to the ground, I know by experience that this phenomenon is not limited to just that particular ball. I have seen many different objects (and I have experienced it with myself) that if let go of above the ground, they return to the ground.

The Scientific Method. Massimiliano Di Ventra, Oxford University Press (2018).
© Massimiliano Di Ventra.
DOI: 10.1093/oso/9780198825623.001.0001

I can then conclude that all these phenomena, which occur with so many distinct objects, share some common thread: no matter their shape, size, color, etc., they all fall if left alone above the ground. It is then natural to think that I could summarize all these observations in some "economical" (preferably mathematical) way.

This *logical synthesis* accomplishes two tasks: it allows me (1) to *describe* what is happening to many, quite distinct material objects, with just one or a few (preferably mathematical) statements, and (2) to *predict* what will happen if I perform experiments with other objects.

Here, I have stressed twice the words "preferably mathematical."

This is because a *mathematical* description is much more powerful than what simple words can accomplish in terms of *describing* phenomena and *predicting* new ones.

(As we will see in the next chapters, prediction of new phenomena is an important and *inalienable* step in the scientific method.)

The mathematical description itself is always *approximate*, and *limited* to a very specific set of phenomena.

Approximate because there is no way to *completely* isolate the particular phenomenon we are trying to describe from the effect of the surrounding "environment" we are not interested in. Hence, the description of such a phenomenon is always *incomplete*.

For instance, when we say that the gravitational force on objects of equal mass on Earth produces the same acceleration, we make a strong approximation: we tacitly neglect the effect of friction that one would naturally have if the experiments were performed in air.

Limited because this description pertains *only* to a particular set of phenomena. For instance, although it would be nice, we do not describe (yet) the full range of gravitational phenomena with the same mathematical description we employ, at the moment, for the quantum ones.

Finally, in order to accomplish the two tasks of *describing* and *predicting*, I need to formulate better the steps that I need to follow, and which make up a *scientific theory*.

Takeaways from this chapter

- We are faced with several *limitations* when we explore the material world.

- One limitation relates to our inability to probe the entire material reality at once: our experimental instruments are limited.

- The (much smaller) part of the whole material reality we study with our instruments is also not fully isolated from the rest we are not interested in.

- Hence, the description of phenomena is always *incomplete*.

- In addition, our (mathematical) *description* of phenomena *only* pertains to a particular (limited) class.

- Typically, different classes of phenomena have quite different descriptions, which cannot always be reconciled.

8

Hypothesis

"Still convinced it's flat?"

So, now that I have observed a phenomenon, say the attraction of objects to the ground, and I want to study it, what do I do?

First, I need to come up with a *hypothesis* or a set of hypotheses, namely *statements*, regarding the objective facts that I observe.

The hypothesis could be as simple as "There is an attraction between any object and the ground," or more general "An attractive force is exerted on and by all material objects in the Universe." Yet other hypotheses may be "All species originate from a common ancestor," "The Universe was confined to a singularity in space and time, long time ago," etc.

The Scientific Method. Massimiliano Di Ventra, Oxford University Press (2018).
© Massimiliano Di Ventra.
DOI: 10.1093/oso/9780198825623.001.0001

The hypotheses need not be mathematically precise or even extremely specific about the objective data. It is enough that they are *descriptive* of what could have generated such data. For instance, "There is an attraction between any object and the ground" is descriptive of the fact that objects left above the ground do fall.

In other words, we say that the hypothesis is *consistent* with what we observe.

But is the hypothesis *true* in the objective sense? Namely, does it share the same level of confidence as the objective facts?

Note that being "consistent" does not necessarily mean "objective."[1] For instance, I could come up with another hypothesis or a set of other hypotheses to describe what I observe.

Since air is everywhere on Earth, I could suggest that it is air molecules which are responsible for the pulling force on the objects. Without additional tests, I could even claim that this new hypothesis is consistent with the observations, since whenever I leave objects above the ground they fall, and invariably I can breathe every time I do the experiments, which means that air is always present when the objects fall.

But then, I could *test* this hypothesis *directly* by redoing my experiments in a vacuum chamber and I would see that air

---

[1] Objective, as intended previously, means "independent of the observer." As I have already discussed, in the context of the hypotheses, there may be a dependence on the observer, in the sense that different observers may come up with *different* hypotheses to account for the *same* facts.

has nothing to do with the fact that the objects always fall to the ground.

The reader may laugh at the weird hypothesis I came up with for gravity and indeed, in this particular case, there is an easy way to check whether it is *consistent* or *inconsistent* with the observations.

But let us now analyze the statement "The Universe was confined to a singularity in space and time a long time ago."

Can I test this statement *directly*? Of course not!

There is *no way* for anyone to go back in time and observe even once, let alone many times, this event. This hypothesis seems to be *consistent* with some observations we make today (like that of background radiation pervading the Universe) and it has been the working hypothesis of modern Cosmology.

But is it a fact? *Not at all*, and despite what many would want us to believe, it will *never* be elevated to the rank of fact, unless we could build a time machine to go back in time and check it, or devise a controlled laboratory experiment in which we reproduce a singularity in space and time, and observe its consequences!

The event itself is beyond our ability to test it *directly*; hence, it is *not* a fact.

What about "All species originate from a common ancestor"? This is again the working hypothesis of modern Biology. But can we test it *directly* and declare it a fact?

Following the scientific method, in order to do so I would again need to go back in time to *when it occurred* (namely, when our first ancestor somehow initiated its transformation into something else) and *test* it by *direct observation*. Alternatively, I need to devise a controlled laboratory experiment in which I do observe such a transformation, say from one species to the one that follows next in the evolutionary path.

Note that, despite what we hear often, it follows from the scientific methodology that adaptation (sometimes called "microevolution") within a single species is not enough to prove that "All species originate from a common ancestor" is a fact. That hypothesis may be *consistent* with the observation of adaptation within a single species, but the latter does not prove *directly* the former.

So the answer is again, no, the above working hypothesis is *not* a fact!

It is the working hypothesis of Biology that attempts to create an economical, unified way of describing what we observe now, but by no means is it an objective fact in the scientific sense.

And I could go on and on with all the other hypotheses that we come up with in the Sciences. They may be *consistent* with the objective data they are meant to describe. However,

> Hypotheses are *not* objective data or facts!

Even the hypothesis considered to be the most obvious of all, namely "An attractive *gravitational force* is shared by all material objects in the Universe," is simply an hypothesis that has allowed us to describe (and, mind, really well!) an enormous amount of data, thus making it perhaps the most consistent of all the hypotheses I have mentioned. But the hypothesis itself is not a fact.

As I have already mentioned, in Einstein's theory of General Relativity there is no concept of a "gravitational force": material objects "interact" with each other through the mutual deformation of spacetime. I put the word "interact" in quotes, because in General Relativity one *interprets* the gravitational interaction between objects as the objects traveling the shortest path in a curved spacetime.

In other words, the word "force" or "interaction" is simply *descriptive* of the observations we make. The instruments we use to make the measurements detect something that we *interpret* as (or call) a "force" or "interaction" or "bending of spacetime."

Such an interpretation could, however, be modified in the future as new phenomena are uncovered. It may well turn out that new data will require extra hypotheses to describe in a consistent way both these new data and the old ones.

For instance, when, and if, we are ever able to describe gravity within the same mathematical structure as Quantum Mechanics , we may very well abandon the hypothesis that there is an attractive force (or bending of spacetime) shared by all objects in the Universe, and replace it with another hypothesis that also takes into account the probabilistic interpretation of atomic phenomena.

Another example is the recent discovery of gravitational waves that has made big news even outside scientific circles.

In the presentation of this discovery, these waves are *believed* to originate from the collision of two black holes. The collision of these black holes has then supposedly deformed the spacetime continuum around them, thus creating the gravitational waves that have been observed in some laboratories on Earth.

Which part of what I just wrote is a fact, and which part is simply a hypothesis consistent with such a fact?

Definitely, the detectors used in the experiments have revealed signals (e.g., the interference pattern of light when a gravitational wave passes by) with an amplitude greater than what can be accounted for by other types of effects (e.g., background noise). What the instruments have measured is a *fact*, so waves have been observed.

The data have been collected, experiments have been repeated, and the results, compared among different teams, have finally been declared objective facts. We have *faith* that the teams of scientists doing this work have all followed the correct procedure as dictated by the scientific method.

What about the *origin* of these waves? Are they really coming from the collision of two black holes some billions of light years away from us? Is this a fact?

No, this is *not* a fact!

Despite the nice *illustrations* of two "black holes" close to each other and "waves" coming out of their collision that we see in magazines or newsfeeds, we have to admit that those are just fruit of the illustrator's imagination.

Black holes cannot even be observed directly. One can infer indirectly their possible presence from their effect on massive objects nearby. So, how in the world does the illustrator even know what they look like?

The back-hole collision is a hypothesis that through mathematical calculations predicts a phenomenon *consistent* with what has been observed by those instruments on Earth.

It is the leading hypothesis now widely accepted by the scientific community, but by no means is it a fact.

Indeed, it could very well be that another hypothesis that is *consistent* with the observations, and yet substantially different from the one of colliding black holes, will turn up (or is already out there but not considered "mainstream").

Mind that if this happened, it would not lessen by a bit the discovery of gravitational waves! It would simply attribute the description of their *origin* to some other possible cause.

From a completely different scientific field, consider the nice pictures of dinosaurs, with their flesh and skin texture beautifully colored, that we see in textbooks or articles.

Are the flesh and skin we see, or even the movements and their vocal sounds "reproduced" in many movies, objective data or facts?

Of course not. Those are again nice illustrations that attempt to put on paper (or on screen) "educated guesses" starting from nothing more than fossils.

The fossils themselves *are* objective data or facts. They are on plain display for all of us willing to visit a museum.

The rest is an "add-on" that may appeal to our eyes and imagination, but nothing more than that.

In fact, in order to prove that a given dinosaur has that particular skin or moves in that particular way, or roars as in the Jurassic Park movies, we would need to go back in time and *observe* all these things *directly*!

Extrapolating fossils into full-fledged (moving and breathing) bodies puts us *beyond* the scientific method and confines us to the realm of pure speculation.

Finally, let me say that although *consistency* of a hypothesis (or a set of hypotheses) with the experiments they are trying to describe is a *necessary* condition, it is *not sufficient* for it to qualify as a *valid* hypothesis.

To explain this important point, we need to move on to what constitutes a *scientific theory*.

Takeaways from this chapter

- Hypotheses are *statements* (not necessarily mathematical) we make to *describe* the observed material world.

- Hypotheses per se are *not* objective data or facts.

- Hypotheses may be *consistent* or *inconsistent* with the observations they are meant to describe.

- It is not always easy to check *directly*, by performing an experiment, whether a hypothesis is (in-)consistent with observations.

- In particular, when hypotheses refer to *past* events they are definitely *not* checkable *directly*, unless we are able to *reproduce* in the lab the phenomena they are meant to describe.

9

Theory

"I find this journal a great source
of inspiration for my theories!"

Once we have settled on one or more hypotheses, our task
as scientists has just begun. As I discussed previously, we
can come up with many hypotheses for the same body
of observations. Some of them are easy to eliminate by
direct measurements, and hence deemed *inconsistent* with
the data, but we could end up with a few competing ones
that superficially all look consistent. How do we distinguish
among these?

We need to come up with a comprehensive set of descriptive statements that allow us to (1) *describe* what we have
already measured, and (2) *make predictions* of new phenomena

The Scientific Method. Massimiliano Di Ventra, Oxford University Press (2018).
© Massimiliano Di Ventra.
DOI: 10.1093/oso/9780198825623.001.0001

yet to be observed. Clearly, these statements need to build logically on the hypotheses we have made.

The set of descriptive statements that accomplishes those two tasks, together with the hypotheses they build upon, is what we call a *theory*.[1]

Note first that, like the hypotheses, the descriptive statements of a theory need not be mathematical. Let us again consider gravity as an example. My theory of gravity could simply be formulated in these two statements: "There is an attraction between any object and the ground" (*hypothesis*) and "Every time I leave an object above the ground to its own devices, it will fall toward the ground" (*prediction*).

These two statements alone are enough to accomplish what I was after: (1) to *describe* in an economical way (namely, with a small amount of hypotheses) all the observations I have made in my lifetime regarding objects left above the ground, and (2) to *predict* that if another observer leaves an object, which I have never thought of, above the ground to its own devices, that object will also fall.

This last step is really *crucial*: with this one prediction, which can be *tested experimentally by independent observers*, I can get rid of other hypotheses that could have been put forward (say, the air one I came up with before).

However, suppose now I want to know more precisely the time it takes a given object to reach the ground once it

---

[1] In certain literature, e.g., in Physics, the word "theory" is sometimes replaced by the word "model." The meaning attributed to both words and to their scientific role is the same, so I will not distinguish between them.

has been released from a given height. This requires further hypotheses, namely that if we could eliminate any possible interference from the environment (say, the air or other obstacles), the "attraction" we guessed could be described by a vector that we call "force" that pulls the object toward the ground (direction of the vector) with a strength proportional to the acceleration of the object (magnitude of the vector).

This statement, usually called the second law of dynamics, is a *mathematical description* that allows us to make incredibly precise predictions on the motion of objects subject to gravity, and not just on Earth. For instance, when applied to planetary objects (and written so that the magnitude of the force between two objects is proportional to the product of their masses, and inversely proportional to the square of their distance) it made it possible to describe the orbits of planets in the Solar System quite accurately, and even the remarkable prediction of the existence of the planet Neptune!

The reader can easily grasp then why Mathematics is the ideal (and universal) language in which to write the descriptive statements of a theory: it allows us to perform accurate, experimental checks of specific predictions that come out of that theory. Since Mathematics is universal, such checks can be planned by anyone on this planet willing to test my predictions.

By limiting the theory to just words (like "objects always fall to the ground") we could not have made, e.g., the prediction of an extra planet in the Solar System, and taken advantage of the predictive power of a mathematically formulated theory in innumerable other cases as well.

It is also important to remember that theories are built upon a *finite* set of hypotheses, which constitutes their foundations.

The finiteness of this set of hypotheses defines the *reach* of a theory in terms of the phenomena that it can *describe* and those it can *predict*.

Two theories built upon *different* sets of hypotheses are *fundamentally* different. I will expand on this point in a subsequent chapter.

For now, I want to make it clear that the fact that a theory (or model) is built on a limited set of hypotheses is not just an abstract concept with only "academic" relevance. It has *practical* consequences as well.

Consider for instance, the hypotheses that go into Quantum Theory (and if you want to extend it to relativistic phenomena, you may consider Quantum Field Theory).

Since this is not a technical textbook, I will not report these hypotheses here. In fact, the only thing we need to know about them is that Quantum (Field) Theory was conceived originally to describe phenomena pertaining to (sub)atomic particles.

However, you could argue that since subatomic particles constitute the building blocks of *all* matter, we should be able to describe *all* phenomena with Quantum Theory, including those that pertain to, say, the *macroscopic* organisms of Biology.

For instance, I should be able to use Quantum Mechanics to describe how two or more organisms interact with each other in their environment.

This "reductionist" approach is often advocated by some physicists. However, while it may have some merit as a "matter of principle," it misses completely the *practical* role of a theory, in this case Quantum Mechanics.

I will stress more this point later in the book using different examples. For now, in order to show how this practical limitation occurs, let us follow this reasoning.

Can we write down the quantum mechanical (Schrödinger's) equation describing *all* the subatomic particles constituting a virus? *In principle*, yes (if we have enough information about the types of atoms making up the structure of a virus).

Can we then write down the same equation for a virus attacking a cell and injecting its RNA into it? Yes, of course. This equation can, *in principle*, be written.

Is this quantum mechanical theory of virus–cell interaction of any help? Absolutely not!

There is *no way* for us to even remotely come close to solving this equation with such a gigantic number of particles, even with the aid of the fastest supercomputers on Earth.

So, while *in principle* we may formulate the interaction of a virus with a cell using Quantum Mechanics, this exercise is of no *practical* value.

Therefore, in addition to the *fundamental* limits imposed on Quantum Mechanics by its own (limited) set of hypotheses, we have also found a *practical* limitation of Quantum Mechanics in tackling some phenomena that, *in principle*, should be within its reach.

Note that I am not saying that no phenomena in Biology fall within the practical description of Quantum

Theory. In fact, some specific effects may even *require* Quantum Mechanics to make a quantitatively predictive description possible, e.g., when chemical reactions occur in some biological complexes due to the interaction of some molecules with light or heat.

However, in all these cases, we *isolate* those (extremely) small parts or molecules (compared to the whole organism) where these reactions occur and study, quantum mechanically, what happens to these small parts (molecules).

We are definitely *not* considering the biological organism in its entirety!

This limitation, however, has no bearing on the applicability of the scientific method, so long as the quantum mechanical description of such a small part of the organism leads to *testable predictions.*

This last point is decisive: we are free to choose whichever subset of the whole system we want to describe (in the present example, some of the organism's molecules), and free to choose whatever theoretical approach we want.

However, to have any meaning in terms of information we can extract from the material world, this description must lead to testable predictions, or it will always be confined (at best) to the realm of hypothesis.

The practical limitation I just discussed is not confined only to Quantum Mechanics, but it is intrinsic to *any* theory (or model) we come up with to describe and predict phenomena.

For instance, referring once more to the Biology example of describing a virus attacking a cell, it would still be impractical to describe all the particles of the virus, the cell, and their liquid environment even if we used classical Newtonian Mechanics. There is no way we could solve Newton's equations of motion for all of the virus's interaction with a cell over the time scales when the virus injects its RNA into the cell nucleus.[2]

We simply do not have the computational resources to do so.

And even if we did have those resources, the question we should ultimately answer is: What *information* do we want to extract from our theory that makes us learn more about this particular phenomenon?

In other words, which *predictions* can we obtain from such an atomic-scale description that cannot be obtained from a simpler theory (model) that simplifies the problem, e.g., by treating the virus and the cell as a system with few parameters and few equations of motion?

As I said, ultimately the choice of what theory (or model) to use to describe a particular phenomenon is left to the researcher, provided the essential role of making predictions out of that theory is not forgotten.

All this should also further clarify the distinction between the *phenomena* that we observe and their *theoretical description*.

First, we observe some phenomenon, say, the falling of objects to the ground when left to their own devices. And this is the phenomenon we want to describe.

---

[2] Note that this method, typically called "classical molecular dynamics," is successfully employed in a variety of fields, such as Materials Science and Chemistry, when the number of particles is small enough to allow for an efficient computation of the equations of motion.

How we describe it is a completely different matter!

I can describe it in simple words, with simple Mathematics from Newtonian Mechanics, or I can use the non-linear equations of General Relativity, or ...

The list of possible theoretical descriptions goes on (and will continue to grow with time). The phenomenon is what it is, while its description may change over time.

By changing the description, however, we introduce new sets of predictions, and if these predictions are tested experimentally, new phenomena may emerge (e.g., bending of light from massive objects as predicted by General Relativity, but completely outside the range of Newtonian Mechanics).

The progress of Science is then a continuous succession of observations, theoretical description of these observations, predictions of new phenomena, and again observations of these new phenomena, if the predictions turn out to be verified by experiments.

We *start* with experiments and we *end* with experiments.

Takeaways from this chapter

- Scientific theories are a set of descriptive *statements* (not necessarily mathematical) that *describe* phenomena *already* observed, and *make predictions* of new phenomena yet to be observed.

- Theories that employ mathematical statements have a greater predictive power than non-mathematical theories.

- Theories need to build *logically* on their foundational hypotheses.

- The number of hypotheses is finite, so that it defines the *reach* of a theory regarding the phenomena it can *describe* and those it can *predict*.

- This is true also in the *practical* application of a theory. Even if a theory could, *in principle*, be applied to a number of different phenomena, unless it allows us to make predictions, it is *practically* useless.

- The reason is that, without predictions, we are left with only a bunch of descriptive statements, whose validity cannot be tested experimentally.

- Experiments are (should be) then both the *beginning* and the *end* of any scientific enterprise.

10

Competing Theories

I have discussed what a scientific theory is. I have stressed that it is not enough that it *describes* the phenomena it was designed to tackle. It also needs to *predict* new phenomena.

Predictions elevate a set of hypotheses from pure speculation to the *testable* construct we call theory.

Predictions also offer the necessary (and testable) checks to distinguish among competing theories.

Suppose I come up with two distinct theories for a given set of observations. Suppose also that those two theories are built on quite different sets of hypotheses.

Which one should we choose?

If the two theories describe the *same* phenomena and lead to the *same* predictions, then there is no specific criterion to decide between the two: *both* theories are scientifically acceptable.

The Scientific Method. Massimiliano Di Ventra, Oxford University Press (2018).
© Massimiliano Di Ventra.
DOI: 10.1093/oso/9780198825623.001.0001

Faced with this situation, which one *should* we choose over the other?

I highlighted the word "should" because, in this case, we need to make a couple of extra considerations.

First, if we are faced with two distinct choices that lead to the same outcome, we tend to follow the "easiest" one. In the case of a scientific theory, this means choosing the one built on the *least number of hypotheses*.

This criterion, sometimes called "Occam's razor," has another obvious advantage: the more hypotheses we introduce to build a theory, the more likely it is that, with time, new experimental data will render one of those hypotheses *inconsistent* with the data.

The second consideration is less edifying for the whole scientific community. Since Science is made by humans, it is not immune from their faults and shortcomings, especially pride and envy.

In recent modern times, these shortcomings have been compounded by the substantial amount of money Science has required to progress, but pride and envy have always been responsible for the major animosities among scientists of any period.[1]

The choice then between competing theories has been, unfortunately, sometimes "settled" by means that have little to do with the logical or mathematical simplicity of

---

[1] See, e.g., the controversy between Newton and Hooke on the theory of colors, or the one between Pasteur and Pouchet on the spontaneous generation of life. The interested reader could learn more about these cases from the book edited by P. Machamer and M. Pera, *Scientific Controversies: Philosophical and Historical Perspectives* (Oxford University Press, 2000).

a theory, and more to do with the "clout" or support a group of individuals receives from the community. When this happens, Science does not necessarily gain.

The situation is definitely easier to sort when the two theories produce at least one or more *different* predictions. In this case, experimental evidence should settle the question, and declare "victorious" the theory that correctly *predicted* the outcome of the experiment.

This example shows once more the central role of experiments or observations and makes it clear that (in an ideal world):

> Science is *not* a democracy: *Nature* rules!

It doesn't matter if almost all the scientists in the world agree on a theory, while an "unknown" chap is the only one with a competing theory. As much as pride can be an obstacle, that majority of scientists must give up their theory (!) if experiments do not confirm it.

Takeaways from this chapter

- If two scientific theories are based on different sets of hypotheses, but make the *same* predictions, there is no obvious criterion to distinguish between them.

- However, if that is the case, we tend to favor the theory based on the *smallest* number of hypotheses (Occam's razor).

- This is because the smaller the number of hypotheses, the less likely it is that, with time, new experimental data will render some of those hypotheses inconsistent with the data.

- Although Occam's razor favors the simplest among two competing theories, the choice between the two can often be dictated by human factors that have nothing to do with the simplicity of a theory.

- If the theories make *different* predictions, then experiments should settle the issue.

11

Can One Theory Be "Derived" from Another?

"Tada!"

I want to make another important point, which superficially may seem just an exercise in scrupulosity, but which should better clarify what a theory is.

We often hear, and we are even taught in some Physics classes, that Newtonian Mechanics can be "derived" from Einstein's theory of Special Relativity as its "limit," when we consider speeds much smaller than the speed of light.

From a purely computational/quantitative point of view, this statement is correct. For instance, if I want to determine the speed of a train with respect to me and use Newtonian

The Scientific Method. Massimiliano Di Ventra, Oxford University Press (2018).
© Massimiliano Di Ventra.
DOI: 10.1093/oso/9780198825623.001.0001

Mechanics to make such calculation, I would obtain practically the same result if I used the equations of Special Relativity in their limit of "low speed."

This is because the speed of light (about 300,000 km/sec) is many orders of magnitude larger than the typical speed of a train (even the fastest trains on Earth cannot go too far beyond speeds of about 400 km/h). Therefore, the ratio between the typical speed of a train and the speed of light can be, for any practical purposes, assumed zero.

Can we then say that Newtonian Mechanics is a "limiting theory" of Einstein's theory of Special Relativity, in situations in which the speeds of the material objects involved are much smaller than the speed of light? It depends on what we mean by "limit of a theory."

Following the above example, we need to recall that there is one important hypothesis that we make in Special Relativity that is *not* part of Newtonian Mechanics: the speed of light is the maximum speed achievable by any material object, and is a constant irrespective of the speed of the light source relative to the observer.

Such a hypothesis is *key* to developing Special Relativity and has allowed us to abandon the notion of an "ether" pervading the Universe and serving as the medium through which light travels.

It is therefore not completely correct to say that one theory (e.g., Newtonian Mechanics) is a "limiting" case of (or is "derived" from) another (e.g., the theory of Special Relativity), because the two have *different* sets of hypotheses, and there is no "limiting procedure" that eliminates one or more hypotheses (e.g., the constant speed of light): either you have them or you don't!

This is true for all theories we construct. For instance, Quantum Mechanics has many additional hypotheses compared to Newtonian Mechanics (or Quantum Electrodynamics compared to Classical Electromagnetism), so the latter one is never truly a limit of the former. Similarly, General Relativity contains a principle (that equates a gravitational field to an accelerating reference frame) that is *not* part of Special Relativity, and so on.

Again, the *calculations* one performs in certain limits (e.g., speeds much smaller than the speed of light) within one theory may agree within certain errors with those obtained from another. However, since they are built on *different hypotheses*, the two theories are nonetheless fundamentally and *conceptually different*.

Takeaways from this chapter

- A scientific theory can never be truly considered as the "limiting case" of another.

- This is because different theories are always built on *different* sets of hypotheses.

- No "limiting procedure" can eliminate one or more hypotheses: either the theory has them or it doesn't.

- This does not mean that some *calculations* performed, and some *predictions* made, with one theory cannot agree with those obtained with another.

- They may actually agree quantitatively within certain limits, even though the two theories are *fundamentally* different.

12

Verifying or Falsifying? And What?

"Don't you think this achievement
puts us out of work?"

Starting from a body of experimental evidence scientists will
move on to construct a theory based on a set of hypotheses
(statements, preferably mathematical). That theory will
make a series of predictions so that it can be . . . what?

We hear a lot about the *falsification* of a theory. The most
outspoken proponent of such a requirement is the late
philosopher Karl Popper.[1]

Falsifiability means that to be declared *scientific* a theory
needs to produce one or more predictions that can be put

---

[1] Karl Popper, *The Logic of Scientific Discovery* (Hutchinson Education,
1959).

The Scientific Method. Massimiliano Di Ventra, Oxford University Press (2018).
© Massimiliano Di Ventra.
DOI: 10.1093/oso/9780198825623.001.0001

to experimental test. If one of these predictions turns out not to be in conformity with the experimental results, then the theory is falsified.

But what does it mean that the experimental results are not in agreement with the prediction(s) of the theory?

Let us go back to the example of discovering the planet Neptune that was predicted by Newtonian Mechanics.

Does this mean that Newtonian Mechanics has been *verified* by that observation?

Not quite. We can conclude from that particular experiment that the hypotheses of the theory, and its set of predictive statements, are *consistent* with the observations to a level that allows us to *describe* not just phenomena on Earth, but also this new discovery at the planetary level.

In other words, we have *expanded the descriptive power* of the theory to a larger set of phenomena. Any other prediction that is successfully tested experimentally does precisely this: *it expands the descriptive range of the theory.*

Although we may loosely use this term, a theory per se *cannot* be truly "verified," because *no* amount of predictions that turn out to be *corroborated* experimentally can give this theory the status of an objective "fact."

Therefore, while the predictions themselves may be *verified* (corroborated) experimentally, the *whole* theory is definitely not.

In other words:

Theories are *not* objective data or facts!

Indeed, from the previous discussion we can summarize the role of a theory simply as follows:

Theories

(i) *describe* a limited set of phenomena, and

(ii) *predict* new phenomena.

Again, the description may not be mathematical, but a mathematical description affords a predictive power that words alone cannot accomplish.

Theories are designed to describe only *specific* phenomena of *specific* natural events. This is the essence of Plato's "to save the phenomena" I mentioned in this book's Preface: it is enough for the theory to be *consistent* with the specific phenomena it is meant to describe, and be able to *predict* new phenomena.

That's all. Nothing else is required from a theory.[2]

When we say that a theory is "universal," what we mean is that it is developed to account for most (if not all) phenomena that could be clustered in a *particular class*. For instance, Newtonian Mechanics has been designed to describe all phenomena related to the motion of classical objects. It was not conceived to describe, say, quantum phenomena.

Even if we could, as nowadays it is hoped for, come up with a single theory that accounts for the (currently known) three forces of Nature (gravity, electroweak, and strong force), that theory would not be a "theory of everything"!

---

[2] See Pierre Duhem's *Essays in the History and Philosophy of Science* (Hackett Pub. Co., 1996) for further discussions on this important point.

It would simply be a theory that *describes* a particular (albeit vast) set of phenomena pertaining to those interactions. It would certainly leave open loads of other questions.

For instance, where does the non-locality of Quantum Mechanics originate from? Are space and time epiphenomena of some other phenomenon? Better yet, other questions and new phenomena will more than likely emerge when new instruments that greatly expand the reach of our senses are devised.

Suppose now that the planet Neptune was discovered but not exactly where the theory predicted it, say some distance away.

Now, should we say the prediction is inaccurate and hence the theory is false?

Not necessarily. We could conclude, for instance, that some other effects have contributed to the discrepancy. Maybe imperfect measurements performed from Earth are the culprit, or the calculations could not be done with very high accuracy due to numerical errors, or maybe some extra effects, not yet accounted for by our theory, need to be considered to have a better agreement with the experiments.

Given the body of other successful predictions, we would certainly not throw away Newtonian Mechanics for this. After all, we did find Neptune!

But again we would not say that we have "verified" such a theory. We would only say that within a certain error (whose origin we may not fully understand yet) the data are consistent with the predictions.

Finally, let us now imagine the worst-case scenario: we did *not* find Neptune, neither where the theory predicted nor anywhere such a planet could be seen within our experimental capabilities.

The prediction did not materialize. Something in the theory is *inconsistent* with the data at a very "critical" level. What exactly should we do about it? Is the theory itself *false*? Of course not!

It only means that we have uncovered a *limit* of the theory, namely that it *cannot describe* certain phenomena that we originally thought it could.

To render this point clearer, let us consider Einstein's incredible prediction (at that time) based on the theory of General Relativity that a ray of light bends near a massive object. This phenomenon was later observed, although Newtonian Mechanics couldn't account for it.

This outcome did not result in scientists tossing out Newtonian Mechanics. In fact, we still apply it very successfully when effects like the one above are negligible. We just uncovered one of its limits.

Similarly, it was found that the speed of light was the limiting speed of all matter. Newtonian Mechanics did not contemplate such a limit.

In fact, absence of such a limit in Newtonian Mechanics may lead to paradoxes due to the superluminal (exceeding the speed of light) transfer of information between objects.

Such paradoxes can be resolved within the theory of Special Relativity or General Relativity, and simply show that Newtonian Mechanics cannot be applied in those contexts.

However, when phenomena involve speeds of objects that are much smaller than the speed of light, we safely employ Newtonian Mechanics to describe such phenomena and make predictions.

Ultimately, it is the experimental test of such predictions that confirms for us whether the choice of Newtonian Mechanics was reasonable to begin with, or whether we have uncovered new effects beyond its reach.

In summary, we do not require the predictions of a theory to "verify" or "falsify" the theory itself, but rather to determine its limits or to distinguish between competing theories.

Takeaways from this chapter

- Theories are *not* objective data or facts.

- They are meant *only* to *describe* phenomena and *predict* new ones.

- If among various predictions of a theory one of them does not materialize (it is *not verified* experimentally), then that theory is *not* "falsified."

- This outcome only means that we have uncovered a *limit* to such a theory; namely, this theory cannot be applied to some phenomena we previously thought it could.

- Even though a *single* theoretical *prediction* can be *verified*, namely *corroborated* experimentally, a theory per se can *never* be "verified," because *no* amount of predictions that turn out to be corroborated experimentally can give that theory the status of an objective "fact."

13

Don't Be a Masochist!

"Don't worry. It's not the boulder coming to us.
It's us going toward the boulder."

Following up on the discussion of the preceding chapter, it is worth noting that although we could use Einstein's theory of General Relativity to describe phenomena that are also within reach of Newtonian Mechanics, this would come at an incredible cost in conceptual effort and computational cumbersomeness.

Newton's equations are much easier to handle than Einstein's equations, and the theoretical framework of Newtonian Mechanics is more intuitive than that of General Relativity when we are faced with phenomena that we experience in our everyday life.

Here, you see again Occam's razor at work: for the set of phenomena where both Newtonian Mechanics and Einstein's General Relativity are equally valid, we choose the

The Scientific Method. Massimiliano Di Ventra, Oxford University Press (2018).
© Massimiliano Di Ventra.
DOI: 10.1093/oso/9780198825623.001.0001

theory that has the least number of hypotheses and provides the simplest description.

To see this "principle" of simplicity at work in an even more compelling way, let us first analyze an example that is frequently misunderstood, not just by lay people, but also within the scientific community.

We often hear the following statements: it is "obvious" that the Ptolemaic system is "wrong" in placing the Earth in a *static* position with the other planets and the Sun moving around it. Instead, the Copernican system is the "correct" one, in which the Sun is *fixed*, with all the other planets, including the Earth, moving around it.

It may come as a surprise to some, but both systems are wrong in some sense, and correct in another.

So, why the apparent contradiction? And, why is the Copernican system the preferred one?

To answer these questions let's start by saying that I, and most of you I am sure, have personally never seen the Earth moving around the Sun (if we do not count science fiction movies or documentaries on the subject).

In fact, most of the time, we observe the motion of the Sun, or any other planet, from the Earth, and in doing so, of course, we are using the latter as the *reference frame*.

On the other hand, when we *imagine* sitting on the Sun ("imagine," because we *cannot* actually do this experiment of relocating to the Sun without dying!), then we use the Sun as the reference frame, and compute the orbits of all the planets, including Earth, from that reference frame.

We can then understand in what sense both the Ptolemaic *and* the Copernican systems are wrong.

If by Ptolemaic system we mean that we choose the Earth as an *absolutely still* reference frame (namely, one that is motionless with respect to everything else in the Universe!), and by Copernican system we mean that the Sun is instead an *absolutely still* reference frame, then we make the same fundamental mistake.

In our physical description, motion of an object has a meaning only *after* we have decided which point of reference to take. In other words, we first choose the reference frame, and only then can we say that an object moves at a given speed, and follows a given trajectory with respect to such a reference frame.

It does not make any sense for us to say that any particular point in the Solar System, or in the entire Universe for that matter, is "absolutely motionless," because the next question we should ask is: with respect to what?[1]

Once this point is settled, we can now understand the (practical) reason for the choice of the Copernican system over the Ptolemaic one.

The Copernican system offers an incredible advantage compared to the Ptolemaic one, *if* we are interested in describing the motion of the planets in our Solar System.

The orbits of the planets as viewed from a reference frame whose origin is the Sun are *simpler* (ellipses) than those we would obtain by describing their motion from the Earth.[2]

---

[1] Also the images taken from spacecrafts or satellites show the motion of the Earth, the Sun, and all the other planets *relative* to (measured from) those spacecrafts or satellites.

[2] To be precise, the Sun is not at the center of the ellipse, but it would be placed at one of its focus points.

That's all. There is nothing more to it! One reference frame allows us, literally, to do calculations on the motion of the planets in a much simpler way than the other.[3]

Now, suppose you want to describe the motion of all the other stars in our galaxy. If you choose the Sun as the origin of the reference frame, the orbits of those stars would be much more complicated than if you chose the center of the galaxy as the reference. And so on.[4]

To exemplify further the fact that we tend to choose the simplest description out of many, equally valid ones, let us use a completely different example. Let us consider a problem from Biology: the motion of a colony of ants as they find food.

This example will also allow me to reconsider the "description limitation" I anticipated when discussing the approximate and limited nature of our explanation of natural phenomena.

You can actually check this example at home if you are willing to have an ant infestation, or, like me, you may have observed it as a kid while playing with these insects.

---

[3] In fact, all of the pictures of the entire Solar System, with all its planets neatly tracing their elliptic orbits around the Sun often found in textbooks, documentaries, etc., are representations of what a hypothetical observer *outside* the Solar System would see were he *at rest* with the Sun.

[4] Also the stellar parallax observed from Earth (namely, the motion of the stars with respect to Earth, when their position is measured at two different days of the year—preferably at six months' distance) is not simply a "direct proof" that the Earth moves around the Sun. It shows the *relative* motion of the stars, including the Sun, and the Earth, as measured, in this case, from the reference frame Earth.

Suppose you are interested in studying the way a colony of ants is able to find food as they come out of their nest.

You will notice that the colony will first send out a few "explorer" ants to search for food in the environment surrounding the nest.

We then observe that if we give them some time, a familiar black path of ants forms, in which ants go back and forth from the nest to a food source, e.g., bread crumbs lying on the floor.

As I mentioned previously, since ants and their environment, including the materials making up their nest, are all made of atomic particles, I could use Quantum Mechanics to study this phenomenon.

I would immediately realize this is a hopeless (some would even say *crazy*) proposition.

My next choice would then be to apply classical Newtonian Mechanics to describe the phenomenon. In this case, I would need to ignore the quantum structure of the atoms in all the ants and their environment and treat those atoms as classical particles following Newton's equations of motion.[5]

I would quickly realize that the number of particles whose motion I need to consider is so gigantic that no available supercomputer on Earth would be able to help me with this.

At this point, I might consider this phenomenon from a different perspective and advance a simpler model *practically* solvable.

---

[5] See Chapter 9, footnote 2.

I would assume (first hypothesis) that the ants use some form of memory (pheromones) to "communicate" and navigate their surroundings.[6] In this way a path from the nest to the food source is chosen by the colony.

I then assume (second hypothesis) that the ants' atomic structure or that of their environment is irrelevant to describe the ants' motion.

After all these approximations, I may therefore model each of them as "structureless" point entities following a very simple equation of motion (with memory) as they travel a network representing the paths they can possibly choose.

This description is *practically* easier to handle than using Quantum Mechanics or Newtonian Mechanics for all the atoms constituting the ants and their environment. I need to solve a considerably smaller number of equations.

Most importantly, it allows me to make predictions that I can test experimentally.

For example, I can ask the question: out of the vast sets of possible paths, which one is chosen by the ants?

Using my model, I predict that if I simulate the following situation, the ants choose the shortest path.

Let us perturb the ants' path *already* formed from the nest to the food source by, e.g., placing somewhere along the path, an obstacle tall enough that now the ants have only *two* possible path choices to move around it (one longer than the other).

---

[6] Memory-dependent motion is indeed a typical feature of physical systems.

If the ants don't lose interest in their food, right after the addition of the above obstacle across their path, they will explore both options (the short and long paths). Given enough time, I would observe that the shortest path is chosen, and a black trail of ants around the obstacle will clearly show it.

I therefore chose the simplest description possible that allows me to make a well-defined prediction.

In addition, I can now evaluate the hypotheses that I made and can conclude that the assumption that ants use a memory mechanism to communicate is *consistent* with what I predicted and observed.

I can also discriminate the hypotheses of my model from other hypotheses that use different mechanisms.

For instance, another theory could claim that ants somehow can visually see the food source directly from their nest with some special optical means. In other words, the ants have a global awareness of their surroundings, a sort of "ant binoculars" that allows them to see beyond any obstacle. In this case, ants would not necessarily choose the shortest path.

Therefore, if another researcher puts forward a *different* description, with very *different* hypotheses, then their alternative theory should at least make the *same* prediction as mine to qualify as *consistent* with the phenomenon observed.

All this to say that even though we have simplified considerably the description of the "ants search problem," this simplification has allowed us to *describe* what we observe and *predict* a new aspect of the phenomenon we study.

This simplification, from the (possibly) more "accurate" ("fundamental"?), but utterly impractical quantum or classical-particle description of either Quantum Mechanics or Newtonian Mechanics, has accomplished the goal we were after: extracting information from the natural world around us.

In summary, we choose a particular descriptive model among competing and equally valid ones according to the *simplicity* with which it allows us to make *predictions* and *measure* (or make calculations about) the phenomena we are interested in, not because one is more "correct" than the other.

You should agree with this point, unless you are a masochist!

Takeaways from this chapter

- The *practical* aspect of a theoretical description cannot be underestimated, since we typically tend to favor the choices that "make our life easier."

- Therefore, the choice we make of a theoretical description, among different and equally valid ones, depends on how *easy* it allows us to make predictions and conceive of new measurements.

- There is no such a thing as a more "accurate" or "fundamental" theory if it does not allow us to compute testable quantities.

14

"Consensus" in Science? What Is That?

"There is an obvious CONSENSUS among scientists about this theory."

We often hear that in this or that scientific community there is a "consensus " on a particular *interpretation* of phenomena or a theory. For instance, there is a clear agreement in Physics that Quantum Mechanics is the best theory to describe and predict phenomena at the atomic or subatomic level.

Does this mean that Quantum Mechanics is "true" or "correct"? Should we even seek to come up with other theories?

Following the exposition in this book, we can certainly say that Quantum Mechanics is, *at the moment*, the best description we have of *a particular class of phenomena*.

The Scientific Method. Massimiliano Di Ventra, Oxford University Press (2018).
© Massimiliano Di Ventra.
DOI: 10.1093/oso/9780198825623.001.0001

However, it would not be a stretch to think that in the future another theory could supersede Quantum Mechanics. Such a theory should be able to describe phenomena that we have observed so far, but also predict new phenomena that the present formulation of Quantum Mechanics is unable to describe.

After all, scientific progress has always happened in this way. For example, a change of scientific "paradigm" (as Thomas Kuhn would call it[1]) could come from the inability of Quantum Mechanics to predict some unexpected phenomenon, which, therefore, falls outside its realm of description. If that were the case, it would require a new foundation; namely, we would need to *add* to or *change* some of its hypotheses.

And even if that were the case, we would not throw away Quantum Mechanics all together. We would still use it for the class of phenomena for which it is valid.

We would simply avoid its application outside of its reach, as we would not use Newtonian Mechanics to describe, e.g., the bending of light from massive objects.

You may be then tempted to think that the scientific community would welcome such a change of paradigm with open arms. Far from it!

Despite their reputation as an open-minded bunch, scientists offer a lot of resistance to change. This is both a blessing and a curse.

It is a blessing, because by resisting a new theory and putting it to strict tests, it allows the scientific community

[1] Thomas Kuhn, *The Structure of Scientific Revolutions* (University of Chicago Press, 1962).

to "filter out" ideas or proposals that do not correspond to objective reality. On the other hand, it is a curse, when it slows down progress to an almost standstill for a while for reasons that are not objectively scientific.

This outcome is particularly felt when a small group of (extremely vocal) scientists promote their own ideas and create their own "circle" of like-minded individuals.

Alternative ideas or theories are then kept at a distance, or worse denigrated. This behavior makes it difficult for newcomers to break into the mainstream of a field of study, even if they bring valuable new suggestions.

I personally think that the situation has become even direr in past decades due to the financial pressures put on scientists, especially the young, untenured ones.

Tenure[2] at research institutions is oftentimes based on the ability of a scientist to receive funding from granting agencies, to publish in "high-impact" journals, to be recognized by their peers, etc.

This compels many young scientists to remain within the boundaries of the "establishment" (or follow the latest "hot fad" of the moment) to avoid making waves that could have a negative impact on their promotion.

This, of course, does not mean that new ideas cannot get through, and finally emerge if they are proven to correspond to reality. It simply means that there is an extra layer of resistance that needs to be shed before that happens.

---

[2] Academic tenure means that the holder cannot be fired for the ideas they support, even "controversial" ones. Clearly, it does not mean they cannot be fired if they commit a felony.

This aspect of "scientific guerrilla" is not just of recent memory.

As an example from the past, consider Boltzmann's kinetic theory of gases that assumed the existence of microscopic objects called atoms and molecules. Boltzmann's contemporaries (especially physicists) were far from thrilled at such a hypothesis. It took further developments and, importantly, several experimental tests, to render the kinetic theory of gases "acceptable" to the wider Physics community.

Other similar examples were referenced in Chapter 10's footnote, and show, as pointed out earlier, that this behavior is part of the fabric of Science. Being a *human* enterprise, it is not exempt from human weaknesses!

Ultimately, however, progress happens and even lonely, isolated voices in the vast cacophony of statements produced by the community will break through. This happens *precisely* because of the existence of an *objective reality beyond us*, which provides the ultimate judgment to this enterprise.

These thoughts can be summarized in a single sentence:

Nothing is *ever settled* in Science!

This sentence holds a very important truth that we should always bear in mind. "Consensus" among scientists *does not* mean that everything is "settled" in a field. This quality of "certainty" is not proper to Science.

When "scientific consensus" is invoked, it often reflects some other "forces" at work in those who pronounce it.

Takeaways from this chapter

- "Consensus" in Science, when it pertains to the *interpretation* of data, does not exist.

- Although data or *facts* may stand the test of time, their *interpretation* is always susceptible to revisions.

- Indeed, *nothing* is ever settled in Science.

- If a group of people claim "consensus" on the interpretation of some data, then typically other "forces" are at play that have nothing to do with Science.

- This "consensus" may have the unwanted outcome of unwittingly slowing down progress.

- Ultimately, however, scientific progress occurs because of the existence of an objective material reality that is the final judge of the validity of a scientific statement.

15

Flow Chart of the Scientific Method

We have now discussed the complete method that is at the core of the Natural Sciences. We can summarize it in a flow chart.

The Scientific Method. Massimiliano Di Ventra, Oxford University Press (2018).
© Massimiliano Di Ventra.
DOI: 10.1093/oso/9780198825623.001.0001

16

The "What" and "Why" Questions

"I welcome questions I cannot answer:
It makes me feel smarter than I am."

With the previous chapters, we completed our discussion of the scientific method. We have seen its object of study and limit of inquiry, so the reader should understand what Science *can* and *cannot* address.

In these final chapters, I will try to reinforce this understanding by addressing questions that we, scientists, are often asked and whose answers, sometimes, are believed to be (incorrectly) the domain of the Natural Sciences.

Some of these questions involve terms that are borrowed from Science, but whose content, in reality, eludes scientific inquiry. Common examples that fall into this category include the following:

The Scientific Method. Massimiliano Di Ventra, Oxford University Press (2018).
© Massimiliano Di Ventra.
DOI: 10.1093/oso/9780198825623.001.0001

What is gravity? *What* is a force? *What* is a particle?

As I have already addressed in this book, such questions are *not*, and *should not be* the domain of investigation of the Natural Sciences. The content of such questions cannot be probed by our senses or extensions of our senses. As such, they are beyond reach of the scientific method.

No doubt, these are questions worth asking. However, they address the *ontological nature* of the subjects under study, namely their very existence and their essence. This is what Philosophy addresses. However, such arguments are not, by their own nature, subject to scientific investigation.

In other words, although the above questions and many statements that employ terms borrowed from Science may be legitimately *logical* and open to *philosophical* discussions, they are nevertheless *not* scientifically testable.

For instance, I could argue that since all material objects in the Universe share an attraction that we call "gravity," irrespective of their mass, or any other properties ("accidents," as philosophers would call them), the *essence* of gravity (what makes it *fundamentally* what it is) is precisely the property of being attractive. Without such an *essential* property of attraction, gravity would not be what it is.

Of course, such an argument is open to philosophical debate. The important point, though, is that it is unquestionably *not* a well-defined scientific statement: I cannot test it experimentally!

While I can measure the strength and direction of what we call gravitational force with very accurate instruments, precisely because my instruments interact with the attributes of material objects (their mass), what *experiment* can I think of that would measure the *essence* of such a reality?

Similarly, we can *measure* with clocks the physical quantity we call time. However, is there a measurement apparatus that can measure its *nature*?

And I could go on and on with *any* material object or phenomenon I can think of. While I can definitely address the material world with empirical means, I cannot use the same means to understand its existence and very essence.

Therefore, there is no point in invoking Science to address the "what questions" above.

These types of questions, however, are not the only ones Science cannot address.

Another set of questions scientists are often asked includes the following:

Why is there gravity? *Why* is there the Universe we observe and not something else? *Why* are there animals or humans at all? Is this the design of an intelligent being (God)?

These "why questions" address the "finality" or "motives" of their respective subjects. As such, they cannot be answered with scientific means.

Note, however, that these "why questions" are *not* the same ones that, say, a biologist would ask regarding, e.g., the "scope" of a given limb in an animal, or the "reason" why certain animals hunt in packs while others do not.

Answers to these questions can be provided within the scientific method, since they point to the description of causes whose ultimate effect is the survival of the individuals and species.

For instance, the act of chewing food with our teeth is related to its easy ingestion, which ultimately helps the nutrients necessary for the survival of our body to be digested more easily. I can then say that the survival of the body is related to the appropriate supply of nutrients, and hypothesize that this supply is facilitated by the chewing of food.

So the question "Why do we have teeth?" could be equally formulated as "What are teeth *for*?" And to this question, I can provide an answer by formulating the hypothesis above, make predictions (e.g., without teeth the food needs to be grinded by some instrument, or I cannot swallow it), test them, etc.

On the other hand, if I asked "*Why* do we want to survive at all?" then I am asking a question that I *cannot* answer scientifically. You cannot just reply, "Because we want to propagate our species." Otherwise, the next question would be "But *why* do we want to propagate our species in the first place?" What is the *ultimate purpose* of propagating our species?

Is there an experiment that I can conceive that answers this question? The answer is no. We cannot think of any experiment that *measures* the reason *why* we, or any other living being on Earth for that matter, want to survive.

Answering this type of question is a task that is best left to *Teleology*: a part of Philosophy that attempts to explain phenomena by their *ultimate purpose*. In fact, by "teleological argument" we oftentimes mean the one for the existence of God from the evidence of order, and, hence, design in Nature.

These questions, again, *cannot* be answered by applying the scientific method, since they do not lend themselves to being testable experimentally. For instance, a physicist cannot answer the question "Why is there gravity and not anything else?"

Science cannot answer this question simply because we do not know, and cannot probe any other alternative, such as a Universe without it.

Can you perform an experiment in which you eliminate gravity in the *whole* Universe? Of course not.

By the same token, arguments that use the "order" in the Universe as strong "evidence" of an Intelligent Designer qualify as religious (or philosophical) arguments, but *have no place* in Science!

In a scientific sense, the word "order" is used in many contexts, such as in the distinction between a solid (the "ordered state of matter") and a liquid or gas ("disordered states of matter").

Does it mean that some things in the Universe are "ordered" (like solids) and some are not? Which parts of the Universe are ordered? What do we mean then that we observe "order" in the Universe? How do we even *quantify* such order for the whole Universe?

And, most importantly, what *predictions* can we extract from that order that are *experimentally* testable?

Another popular argument in favor of "design" involves the concept of "fine-tuning" of the "fundamental" physical constants, such as the fine-structure constant, or the charge of the electron, or the strength of the nuclear force.

Again, without entering into worthy debate as to whether there is a Designer, the argument of "fine-tuning" is beyond our experimental reach.

In fact, "thought exercises" like "What happens if some of what we call 'fundamental' or 'universal' physical constants are changed even by a bit?" do not stand the scientific test. There is no way for us to do that experiment and *observe* its consequences!

At most, we can provide an "educated guess" as to what would happen if one of those constants, say, the speed of light in the vacuum, were not what it is, thus imagining "worlds" where superluminal information is transmitted. (Writers of science fiction novels or movies do this all the time.)

However, excluding science fiction, we cannot change the value of the speed of light in the vacuum, or of any other "fundamental" constant for that matter.[1] We cannot even conceive of changing one of them while keeping the others fixed, and test experimentally such a change.

In summary, "understanding" in Science is not the same as "understanding" in Philosophy or Religion. In Science we *describe* relations between different material objects and their relation with us, the observers. We cannot understand *why* they exist or *what* their ultimate purpose is.

Takeaways from this chapter

- Science *cannot* answer questions about the *existence* and *essence* of the material reality ("what questions") and its *ultimate goal* ("why questions").

[1] Note that I have added the words "in the vacuum" on purpose. The speed of light *does* change in materials other than the vacuum. It is the speed of light in the vacuum that is considered a "universal constant," irrespective of the relative speed of the observer and the light source.

- The "what questions" are *ontological*; the "why questions" are *teleological*.

- Therefore, they pertain to other sources of knowledge, such as Philosophy or Religion.

- We cannot conceive of any experiment that allows an *experimental test* of the above questions.

- Attempts to answer those questions only *within* the boundaries of Science will undoubtedly lead to *illogical* and *false* conclusions.

17

"Scientism": Abusing the Scientific Method

"This idea was dead on arrival."

I started this book by quoting a sentence that I proved to be illogical: "Science is the *only* way we know truth."

Although this sentence refutes itself, it is still used in conversations, and thus shows an alarming pattern among some scientists, and other people who use "scientific arguments" for their own agendas, or to defend their views on Religion or Philosophy.

Such arguments tend to polarize people against each other, rather than bringing to the table an honest and balanced discussion.

The Scientific Method. Massimiliano Di Ventra, Oxford University Press (2018).
© Massimiliano Di Ventra.
DOI: 10.1093/oso/9780198825623.001.0001

For example, we hear sentences like "You must be *insane* or *uneducated* if you do not *believe* in . . ." (pick your favorite scientific subject).

These sentences have two major problems.

For one, they provide *ad hominem* attacks, namely attacks directed at the person(s), rather than preparing the interlocutors for an open discussion about the scientific subject under review.

Secondly, they use the word "believe" as if what we are discussing is a "divine" or "revealed" truth. Argumentations such as these are *not* and *should not* be part of any serious discussion about Science.

They are indeed an abuse of the scientific method to its core since, as we have seen in this book, there is no "revealed truth" in Science. The only objective *facts* are those that can be observed with our senses or extensions of our senses. Nothing else is a fact in Science, and no amount of "consensus" among scientists can change that, or "settle" a scientific inquiry!

Applying the "principle" that Science is the only way we know truth leads also to arguments that attempt to explain the "origin of the Universe."

For instance, we hear opinions from some scientists that the Universe could have "emerged" from a vacuum (quantum) fluctuation. Therefore, they conclude that there is no need of a Creator, since a vacuum fluctuation can create "something from nothing."

It should be now evident to the reader that this is also an abuse of Science since it utterly confuses a philosophical/logical argument with a scientific one.

First, let us note that a quantum fluctuation is *described* in Quantum Field Theory as an event that occurs at a given *time* and at a specific point in *space*, and involves two conjugated *particles*, e.g., an electron and a positron, for a very short time.

As such, a quantum fluctuation *already* involves "something," namely the *existence* of time, space, matter, charge, etc.

On the other hand, as I have already mentioned, "creation *ex nihilo*," namely creation from "nothing" (truly the absence of *any* created thing), is properly discussed in Philosophy and Religion. It is the very act of creating *everything, including* time, space, matter, etc.!

In other words, in Philosophy and Religion, "nothing" means really *absence of anything*, including realities such as space and time, that in Science we must assume as a given, to even *begin* our description of material phenomena.

This would be true whether the Universe were *finite* in time and space, or *infinite* in both. Even the latter case (a spatially infinite Universe that always existed) would not avoid the need to question who or what has put *into existence* the very realities of space, time, matter, etc.

The event of creation itself *cannot be probed experimentally*: it is beyond scientific reach. You may choose to *believe* it or not, but that belief is something that pertains to Religion, not Science.

Finally, the assumption that any scientific statement contains truth in itself may lead to the elevation of mere hypotheses to "theories" or worse "facts."

A typical example is the "hypothesis" of a "multiverse," namely the existence of many (infinite?) "universes" apart from ours.

Despite the very appealing illustrations we see in some magazines or news articles about "multiverses," this concept might qualify *at best* as a "hypothesis," but it fails to *predict* any consequences thereof that are testable. Hence, it is again beyond the limit of scientific inquiry.

If there are other universes beyond ours and we cannot even communicate with them, how can we *measure* any of their physical properties?

And if they are connected to ours, so that, in principle, we can send signals from ours to the others, isn't this simply a single, unique Universe? If so, the question should be "How do we *measure* the physical presence of these different 'parts' of our Universe?"

Lest the reader think I am only blaming this abuse of Science on those who promote a certain worldview, I stress again that the abuse of the scientific method is also perpetrated to reach diametrically opposite conclusions.

For instance, let us consider once more the claim that since there is "order" in the Universe, then this is "proof" of the existence of an "Intelligent Designer." Such a statement, again, may find its natural place in Religion or Philosophy, but it has no scientific value.

To address this topic by scientific means would require a definition, preferably mathematical, of such an "order," an objective description of phenomena in terms of that "order," and, finally, testable predictions based on it.

And even if all this could be accomplished, it would still not answer the question of who "ordered" such a material reality, or who has created it *ex nihilo*. We may answer these questions using philosophical, logical, or religious arguments, but definitely not scientific (empirical) ones.

Similarly, we hear that since the Big Bang "occurred," it is a direct "proof" that an act of creation happened.

As I have discussed previously, (1) we did not make any *direct* observation that shows the Big Bang actually occurred. It is a hypothesis *consistent* with some observations, but not a fact in itself.

(2) Even if this event did occur, we again face the same scientific difficulty in transitioning from the *description* of the initial stages of the Universe to the creation *ex nihilo* of such a Universe.

Going from a singularity in spacetime to *creation* of that space, time, matter, etc. is a far cry and does not follow from *any* scientific argument.

Finally, a similar argument that uses the "complexity" of the natural world as evidence of a Creator has been made. In the same vein as the earlier discussion about order, "complexity" (which also in this case would require a precise mathematical definition leading to testable *predictions*) cannot be adduced as a "scientific statement" to validate a religious argument. While there could be a logical relation between the two, it is not properly a scientific one.

For instance, "complexity" in Physics is often used to describe situations in which a physical system has many parts ("degrees of freedom") interacting with each other. This large number of elements then gives rise to some phenomena that cannot be easily described as originating from the physical properties of the single elements themselves.

In simple terms, a "complex" physical system is "more than the sum of its parts."

However, what I just wrote is still within the confines of the scientific method. I can certainly describe (even mathematically) these complex systems, make predictions about their properties, and finally test such predictions.

How do I go from the *description* of these complex physical systems to who or what has *created* them, if empirical means are the only ones at my disposal? The *only* way I can answer this question is with philosophical or religious arguments.

In order to do so, I can *definitely* make a logical or philosophical step, but *not* a scientific one. Science has nothing to offer here.

In conclusion, I hope I have conveyed to the reader that abusing the scientific method beyond its empirical reach, namely beyond what we can measure (I call this abuse "Scientism"), can lead to *non*-scientific conclusions.

This does not mean those conclusions are illogical (or irrational) or cannot otherwise be proven by a solid, logical argument. In fact, they may have their rightful place in Philosophy or Religion. But certainly not in Science.

Takeaways from this chapter

- *Scientism* is *any* abuse of the scientific method for purposes and in contexts that are *not* empirical.

- Scientism is most often used, although not always, to reach diametrically different conclusions about Religion.

- It is seriously *damaging* to our ability to reason logically.

- Rather than acknowledging the strengths and opportunities offered by *all* sources of knowledge, Scientism diminishes, often ridicules, and damages their role.

- It is itself a very bad form of "religion" that should be rejected *unequivocally* by laymen and practitioners alike.

- Science has *no use* for this sort of "religion," and we all would do well to guard against it.

18

Final Thoughts

"No doubt your presentation has
been a success!"

It is time to recap what we have discussed in this book and
offer thoughts on some points that may still require further
discussion.

We have seen that the *only* object of study of the Natural
Sciences is the *unique* material reality that is accessible via our
senses and extensions of our senses (experimental probes).
In the scientific context, the *existence* of such a reality is a *logical necessity*: this statement cannot be probed experimentally.

As scientists, we *must* agree on the existence of such a
unique reality, or nothing we do makes any sense. Just imagine if two scientists making experiments on the *same* phenomenon found *different* results and concluded that one

The Scientific Method. Massimiliano Di Ventra, Oxford University Press (2018).
© Massimiliano Di Ventra.
DOI: 10.1093/oso/9780198825623.001.0001

must live in one reality and the other in another … That would make it utterly impossible to conclude anything on the phenomenon they are studying.

Moreover, the *only* knowledge Science can provide of this material reality is the *description* of the interactions between the different material objects among themselves as probed experimentally by the observers.

Science *does not* and *cannot* answer the question of *what* is the essence of such a reality (ontological question) or *why* there is such reality (teleological question). These are valid questions that belong to Philosophy or Religion, not Science.

The scientific *description* attempts to create a logical *synthesis* of a particular set (not all) of the data (objective *facts*) and phenomena that we observe, by making assumptions (statements), which we call *hypotheses*, to build *theories*.

These theories *cannot* be only *descriptive*: to have any scientific value they need to be *predictive* of new phenomena. Otherwise, they are, at best, hypotheses, or, at worst, fruits of our imagination.

Theories, like most of their hypotheses, are *not facts*! They are a set of statements used to build a *description* of specific phenomena and *predict* new ones. Theories can, and will, change as new phenomena are uncovered.

This is it! This is the whole goal of Science and its limit.

Any abuse of this methodology, such as the use of scientific arguments in contexts that are not empirical, like the proof or disproof of a Creator, is not and should not be part of Science.

Another type of abuse might also include the recent suggestion to change the scientific method in order to accommodate certain theories that do not make experimental predictions.[1]

In reply to this proposal, we may simply ask: How do we *objectively* judge the value of such theories if there is no *material reality* we can compare them with?

In fact, while we may argue (as I will do in a moment) about the timeframe within which predictions could be tested, we cannot debate about the *necessity* of such tests!

It would be as though two chess players were debating whether the ultimate goal of the game is to checkmate the adversary king. They may debate whether to play a short or long game, but not on the final goal of the game itself.

Concerning the timeframe necessary to test predictions, one of the open questions is what to make of some theories that make predictions that cannot be tested, not just during our lifetime, but possibly in several lifetimes.

We know, for instance, that an observation consistent with what we call "gravitational waves" has been recently made.

The prediction has been tested but almost 100 years after the publication of the original idea. If we lived during Einstein's time, we would have thought such a prediction so far out into the future to be almost impossible to test. Nonetheless, we did.

How do we regard then those theories that make predictions of phenomena at high energies or length scales that are impossible to attain with present instruments?

---

[1] See, for example, Richard Dawid, *String Theory and the Scientific Method* (Cambridge University Press, 2013).

It is difficult to see how we could improve our instruments to reach such tiny length scales or high energies within a century, let alone within a few decades. Do those predictions qualify as valid?

Following the scientific methodology, the answer is yes, but with a provision. They are valid predictions if they follow logically from a set of hypotheses, and if those hypotheses can also be used to describe a class of phenomena that we have *already* observed.

Consider again the prediction of gravitational waves, made after Einstein proposed his theory of General Relativity. The latter could already describe accurately the anomalous precession of the perihelion of the planet Mercury. The other predictions, such as the bending of light due to a massive object, gravitational redshift, and many others, including gravitational waves, were all tested subsequently.

Nonetheless, Einstein's theory *already* described a class of phenomena *previously* observed. It was just a matter of time to test its other predictions and see whether they would be corroborated (hence, expanding the reach of the theory) or not (thus, determining its limits).

The problem arises if theories, despite not having been experimentally tested yet, are elevated to "accepted theories" or, worse, "facts."

There is truly no point in claiming that such theories are "accepted" by the scientific community, since their scientific value has not been really proved yet by its true, ultimate judge: Nature.

As I have discussed in this book, the descriptive power of a theory should not be construed as a "fact."

To all of this let me add yet another point.

In some fields of the Natural Sciences, Physics in particular, there is also a tendency among theorists to put an exaggerated emphasis on the mathematical complexity and "intricacies" of a theory. It is as if the more "complex" the theory is, the more "important" it must be.

This attitude completely misses the ultimate goal of a theory: to *describe* phenomena and *predict* new ones, *irrespective* of its mathematical "intricacies."

Of course, if new types of Mathematics, as difficult as they may appear at first, help in the prediction of new phenomena, then they are more than welcome to enter our toolbox of useful mathematical tools.

The Mathematics of Hilbert operators in Quantum Mechanics or differential geometry in General Relativity are obvious examples.

In addition, it might happen that a different, oftentimes more "economical" (simpler) *mathematical* formulation of a theory *inspires* new directions of study. This is all good for the progress of Science.

However, a problem may arise if Mathematics is elevated to the *only guiding principle* to discover new phenomena. This is because doing Mathematics *for its own sake*, while employing it in the Natural Sciences, may lead to logically derived mathematical constructs that, however, have little or no bearing on natural phenomena. In fact, by following only logical, mathematical steps while building a theory may even lead to completely unphysical results.

For instance, an infinite quantity is a well-defined concept in Mathematics, but it is not physically possible in the Natural Sciences: we cannot *measure* infinities because our instruments are finite.

In general, quantities firmly grounded upon the foundational axioms of Mathematics may have no correspondence in the material world.

Therefore, while we must surely value Mathematics as an incredibly powerful tool, it is important to guard from overemphasizing its importance in the Natural Sciences, whose ultimate goal is to describe the *natural* world, irrespective of what mathematical description we employ, if we employ one at all.

In addition, a theory should be preferred to another only if it allows describing a certain class of phenomena *more easily* (Occam's razor), or extends its reach to *other phenomena* that the other theory cannot describe or predict.

If neither of these two criteria hold, namely the theory is far too "complex" (in a mathematical sense) to allow calculations and does not make any other prediction, in addition to those offered by other existing theories, then there is no reason in elevating such a theory to the rank of "accepted theory."

If this happens, it is typically because some individuals have exerted some type of "pressure" on the community to promote their own agenda.

When this happens, Science does not always gain.

I am discussing these points because both the number of scientists and the scientific output is steadily increasing.

For this reason, it is imperative to discuss what reasonable limits we should impose on what someone can say regarding the material world. Otherwise, anything goes.

This is why the past century has seen the refereeing system (namely anonymous peer reviews of scientists' work) take an increasingly important and central role in judging new scientific claims, with regard to both research output and the award of funding.

Although such a system is not infallible,[2] and definitely could be improved, it does provide some needed accountability on the work of scientists.

However, the discussion of how to improve it, or how to come up with alternative forms of "checks and balances" is still ongoing. Any suggestion is therefore welcome.

Ending on a positive note, I am convinced that if the scientific method is applied with intellectual honesty, it will allow us to make sense of the facts that Nature will reveal in the process of discovery.

This is precisely *because* the objective reality we study is there, *irrespective* of the fads of the moment, or the opinions of a few.

By keeping this in mind, we can certainly say that humanity will always benefit from this marvelous enterprise we call Science.

[2] For instance, the refereeing system could be manipulated by a small number of scientists promoting their own ideas and the ideas of their "trusted circle," or, worse, by some editors interested only in increasing the popularity of their journal.

Further Readings

Pierre Duhem, *Essays in the History and Philosophy of Science* (Hackett Pub. Co., 1996).

In my view, this book more than any other clearly reflects on the limits and possibilities of Science. It was written by a well-known physicist at the beginning of the past century, but it is still very valid to this day. A very enjoyable read that expands on what I have discussed in my book.

Thomas Kuhn, *The Structure of Scientific Revolutions* (University of Chicago Press, 1962).

This book was written by a philosopher who trained as a physicist. It focuses mostly on the "sociology" of Science, and how scientific revolutions (changes in scientific "paradigms," as he calls them) have occurred historically. It is a nice compendium and an easy read, but does not address the method as clearly and in depth as Duhem's book.

Karl Popper, *The Logic of Scientific Discovery* (Routledge, 1959).

This is a book on epistemology written by a well-known philosopher of Science. One of the main tenets of Popper's work is that scientific methodology should always lead to *falsifiability* of a scientific theory. He claimed this is because no amount of experiments could ever "prove" a theory, but a single one is enough to contradict it.

As I have discussed, such eventuality does not mean that the whole theory is false. An experiment that turns out to falsify a *prediction* of a theory simply sets new boundaries for the theory, but by no means invalidates all its tenets.

Scientific theories are not *mathematical conjectures*, which can be proved false by a single counter example. This is because the goal of Natural Sciences is to *describe* an objective *material* reality and *predict* new phenomena, whereas Mathematics *extracts* conclusions that follow logically and unequivocally from its foundational axioms, irrespective of whether they correspond to a material reality.

About the Author

Massimiliano Di Ventra is Professor of Physics at the University of California, San Diego. He has published the textbook *Introduction to Nanoscale Science and Technology* (Springer, 2004) for undergraduate students, and the graduate-level textbook *Electrical Transport in Nanoscale Systems* (Cambridge University Press, 2008).

Illustrations by Matteo Di Ventra ($\mathcal{MD}^2$).

Index